Reliability-Based Design of Engineered Wood Structures

NATO ASI Series

Advanced Science Institutes Series

A Series presenting the results of activities sponsored by the NATO Science Committee, which aims at the dissemination of advanced scientific and technological knowledge, with a view to strengthening links between scientific communities.

The Series is published by an international board of publishers in conjunction with the NATO Scientific Affairs Division

A Life Sciences **B Physics**	Plenum Publishing Corporation London and New York
C Mathematical and Physical Sciences **D Behavioural and Social Sciences** **E Applied Sciences**	Kluwer Academic Publishers Dordrecht, Boston and London
F Computer and Systems Sciences **G Ecological Sciences** **H Cell Biology** **I Global Environmental Change**	Springer-Verlag Berlin, Heidelberg, New York, London, Paris and Tokyo

NATO-PCO-DATA BASE

The electronic index to the NATO ASI Series provides full bibliographical references (with keywords and/or abstracts) to more than 30000 contributions from international scientists published in all sections of the NATO ASI Series.
Access to the NATO-PCO-DATA BASE is possible in two ways:

– via online FILE 128 (NATO-PCO-DATA BASE) hosted by ESRIN, Via Galileo Galilei, I-00044 Frascati, Italy.

– via CD-ROM "NATO-PCO-DATA BASE" with user-friendly retrieval software in English, French and German (© WTV GmbH and DATAWARE Technologies Inc. 1989).

The CD-ROM can be ordered through any member of the Board of Publishers or through NATO-PCO, Overijse, Belgium.

Series E: Applied Sciences - Vol. 215

Reliability-Based Design of Engineered Wood Structures

edited by

Jozsef Bodig

Engineering Data Management, Inc.,
Fort Collins, Colorado, U.S.A.

Springer-Science+Business Media, B.V.

Proceedings of the NATO Advanced Research Workshop on
Reliability-Based Design of Engineered Wood Structures
Florence, Italy
2–5 June 1991

Library of Congress Cataloging-in-Publication Data

Reliability-based design of engineered wood structures / edited by
Jozsef Bodig.
p. cm. -- (NATO ASI series. Series E, Applied sciences ; vol.
215)
Papers originally presented at an advanced research workshop.
Includes index.
ISBN 978-90-481-4168-5 (HB : acid free paper)
1. Building, Wooden--Congresses. 2. Reliability (Engineering)-
-Congresses. 3. Building, Wooden--Standards--Congresses.
I. Bodig, Jozsef. II. Series: NATO ASI series. Series E, Applied
sciences ; no. 215.
TA666.R45 1992
624.1'84--dc20 92-15408

ISBN 978-90-481-4168-5 ISBN 978-94-015-8044-1 (eBook)
DOI: 10.1007/978-94-015-8044-1

Table of Contents

PAPERS PRESENTED (continued):

ACKNOWLEDGEMENT

Acknowledgement is given to the Scientific Affairs Division of NATO and the U.S. National Science Foundation for co-sponsoring the ARW. Further acknowledgement is given to the U.S. National Forest Products Association for sponsoring a number of the workshop participants. A special appreciation is extended to the members of the Organizing Committee: Drs. J. David Barrett, Peter Glos, Hans Jorgen Larsen, and Robert Leicester for their efforts in making the workshop a reality.

A sincere expression of thanks is due to Professor Ario Ceccotti of the Civil Engineering Department, University of Florence. Dr. Ceccotti and his assistants, Ms. Gabriella Montagnani and Ms. Simonetta Vanacore, provided invaluable assistance in making the local arrangements.

Appreciation is also expressed to the staff of Engineering Data Management, Inc., specifically Mr. Ronald W. Anthony, Assistant to the ARW Director, and Ms. Lisa Nelson, Administrative Assistant to the ARW Director, for their work on the many details associated with the organization of the workshop and publication of the proceedings.

INTRODUCTION

Reliability-based design (RBD) procedures for engineered structures are being developed and quickly gaining acceptance by code agencies throughout the world. Numerous organizations are involved in the development of national or regional codes without the benefit of interchange of ideas and methodologies. Harmonization and coordination of these activities is absolutely essential if the ever-increasing international commerce is to flourish.

This NATO Advanced Research Workshop (ARW) was organized to bring together, for the first time, experts on RBD and related subjects from various countries to assess the current knowledge and recommend new developments. Further, due to their unique nature and great economic significance in most parts of the world, special emphasis was placed on engineered wood structures. For example, in North America more wood products are used in construction than all other materials (steel, concrete, brick, etc.) combined. However, the wood industry segment, historically, receives less attention and smaller financial support for new developments than other construction materials.

Similar, but largely independent, RBD developments are being conducted in Europe, North America, New Zealand and Australia. Experts from these regions were brought together to exchange information on current work, propose new developments and to provide means of international coordination. Thus, this ARW provided an opportunity to advance the cause of RBD of engineered wood structures.

The objectives of the ARW were to review the state-of-the-art on RBD procedures; exchange ideas and approaches used with various construction materials; identify limitations of current methodologies; and define needed developments toward the enhancement of RBD with special emphasis on internationally harmonized codes for wood structures.

To aid in accomplishing the above objectives, the following topics and corresponding sub-topics were selected for detailed examination:

A. FUNDAMENTALS OF RELIABILITY ASSESSMENT

- Structural reliability theories
- Structural reliability analysis
- Calibration
- Strength limit states
- Reliability philosophies for wood structures

B. RELIABILITY ASSESSMENT OF MULTI-MEMBER STRUCTURES

- Criteria of evaluation
- Multiple limit states

J. Bodig (ed.), Reliability-Based Design of Engineered Wood Structures, 1–2.

- Interaction of components in built-up elements
- Interaction of elements in whole structures
- Modeling of uncertainties in whole structures
- Unique behavior of wood structures

C. MATERIAL RESISTANCE CONSIDERATIONS

- Statistical treatment of material resistance
- Uncertainties in the assignment of nominal resistance
- Resistance assignment through imperfect correlation
- Role of proof loading, nondestructive evaluation, and quality assurance
- Special features of wood products (duration of load, influencing factors, environmental degradation)

D. FORMULATION OF INTERNATIONALLY HARMONIZED CODES

- Harmonization of nominal resistance assignment
- Harmonization of load and resistance factor assignment
- Harmonization of material testing methods
- Current status of national and regional RBD codes
- Strategies needed for harmonization and code format selection of wood structure codes
- Mechanism needed for harmonization of wood structure codes

The activities of the ARW were focused on the four topics. In addition to an overview by a keynote speaker, state-of-the-art presentations were given by seven participants. These presentations formed the basis of the deliberations of the four working groups.

Each of the working groups was charged with addressing the following issues:

- Current knowledge
- Limitation of the current knowledge
- Needed research and development
- How to generate the needed research
- Mechanism for accomplishing the needed developments

It is hoped that this ARW will signal the beginning of a long-term activity toward the exchange of information; development of common nomenclature; coordination of materials evaluation methods; and the harmonization of engineered timber design codes throughout the world.

Jozsef Bodig, Ph.D.
Director, NATO ARW

STATUS OF RELIABILITY-BASED DESIGN IN NORTH AMERICA: IMPACT FOR ENGINEERED WOOD CONSTRUCTION

By

Bruce R. Ellingwood
Department of Civil Engineering
The Johns Hopkins University
Baltimore, MD 21218 USA

ABSTRACT. Safety and serviceability requirements for structural design in North America and the role of probability and statistics in addressing uncertainty are described in a historical context. Load criteria and performance requirements for limit states design are developed using reliability analysis techniques. The reliability analysis integrates statistical data on loads and strengths to yield criteria that have a conventional appearance and are consistent with prescribed measures of reliability. The ongoing development of load and resistance factor design (LRFD) criteria for engineered wood construction requires special attention to certain characteristics of wood products, such as their inherent large strength variability and the effect of duration of load on the design strength.

1. INTRODUCTION

Codes and standards used in structural design contain provisions to ensure that the structure is safe under extreme loads and remains functional under service loads. These provisions include sets of equations such as (AISC, 1986; ASCE, 1990),

$$\text{Design strength} \geq \text{Effect of design loads} \quad (1)$$

$$\text{Deflection} \geq \text{Effect of service load} \quad (2)$$

The basis for these safety and serviceability checks and the manner in which engineers perform them has changed considerably during the past 40 years. During this period of time, we have seen the adoption of limit states design, the introduction of the computer as a tool for routine analysis and design, and an increasing acceptance of probabilistic methods for dealing with uncertainty.

2. EVOLUTION OF SAFETY CHECKING

Structural design codes historically have been based on professional experience, judgement and intuition. Despite recent advances in computational mechanics, structural loads, strengths, and their complex interrelationships remain difficult to predict accurately. Uncertainties arise from inherent variabilities in loads and material strengths, dimensions, and natural and man-made hazards, as well as

J. Bodig (ed.), Reliability-Based Design of Engineered Wood Structures, 3–19.

from insufficient knowledge and human error in design and construction. Uncertainty in all phases of the building process is a fact of life, and prudent engineers and builders usually have taken conservative approaches to construction. In the 19th century, along with the advent of formal structural calculations as an integral part of design came the use of factors of safety aimed at ensuring safe and serviceable performance.

These early design methods were based on the notion of elastic behavior and took the form of what we now refer to as allowable stress design (ASD). The basic idea in ASD was (and is) to select the loads conservatively, calculate the stresses from these loads by elastic analysis, and check that these stresses are less than some safe fraction of the limiting stress at which failure occurs in yielding, fracture, buckling, etc. In equation form, we have,

$$f_k \leq F_k/FS \qquad (3)$$

in which f_k is the stress due to the applied loads, F_k is the limit stress and FS is the factor of safety. The effect of uncertainties in strength and loads are combined in this one FS. Factors of safety have declined more or less continuously since the late 19th century. In the 1880's, for example, the allowable stress for steel members in tension was $0.4F_y$, implying FS = 2.5. By the 1940's, the FS had decreased to abou 1.67 ($F_t = 0.6F_y$ for tension), where it has remained. The safety factor was a judgmental reflection of the confidence placed in existing analysis and design methods. Occasionally, of course, overconfidence led to failures, and the design procedures were made more conservative.

In the late 1940's and early 1950's, the late Professor Freudenthal published several (now classical) papers, observing that many load and strength variables exhibited statistical regularity and that uncertainty could be described quantitatively by probability distributions. He suggested that the failure probability, P_f, should replace the judgmental FS as a quantitative measure of safety and performance. Freudenthal's research laid the groundwork for the development of the field of structural reliability.

American Concrete Institute Committee 318 introduced ultimate strength design as an alternative to allowable stress design in the 1963 edition of ACI Standard 318. This move included separating the FS into load factors and resistance factors to account, respectively, for variabilities in loads and in strengths. During the preparation of ACI Standard 318-63 in the early 1960's, Committee 318 attempted to set the load and resistance factors so that the probability of exceeding the factored gravity loads was approximately 10^{-3}, while the probability that the actua strength would be less than factored strength would be about 10^{-2}. Formal reliability methods were not used to set the resistance and load factors, however. It was felt (naively, with hindsight) that this would ensure a member failure probability of about 10^{-5}.

In 1969, the American Iron and Steel Institute and the American Institute of Steel Construction initiated research to develop a practical limit states design specification for steel structures using probabilistic methods. This specification used a format referred to as load and resistance factor design (LRFD). The LRFD Specification finally was published in 1986, 17 years after work had begun (AISC, 1986). In 1990, AISI completed a draft specification for cold-formed steel construction.

For codification purposes and facility in design, it is desirable to uncouple the load and resistance sides of Eq. (1) and to make the load requirements - the right hand side of Eq. (1) - independent of construction material. This avoids the confusion that would ensue if different construction material specifications developed their own load requirements and facilitates design where construction materials are mixed in a structure. Work was initiated in 1979 to develop uniform load factors and load combinations that could be included in American National Standard A58, Minimum Design loads for Buildings and Other Structures. The load factors and load combinations were determined through a code optimization procedure (Ellingwood, et al, 1982). They appear in ASCE Standard 7-88 (ASCE, 1990), and have been adopted by both the AISC and AISI specifications. A partial set of these combinations is (ASCE, 1990):

$$1.2\,D_n + 1.6\,L_n \tag{4}$$

$$1.2\,D_n + 1.6\,S_n + 0.5\,L_n \tag{5}$$

$$1.2\,D_n + 1.3\,W_n + 0.5\,L_n \tag{6}$$

in which D_n, L_n, S_n and W_n are nominal dead, live, snow and wind loads. ASCE Standard 7-88 contains load combinations for earthquake as well; however, earthquake-resistant design requirements are currently in a state of flux and it is likely that this load combination will be changed in the near future. No additional information during the past decade has suggested that changes in Eqs.(4) through (6) are warranted. Structural analysis using these load combinations defines the required strength - the right hand side of Eq. (1).

Engineers are influenced in their selection of structural materials by economic constraints, material availability, and confidence in their reliability. Wood as a construction material scores well in the first two categories, but not in the third, especially now that steel and reinforced concrete have moved toward limit states design. Wood presents a number of problems as an engineered material that are not encountered for steel and concrete, at least not to a significant degree. Wood is a natural material with growth defects and relatively large inherent variability in its strength and stiffness. Classifying wood products for structural usage by species and grade only partly reduces this variability. Engineered wood construction encompasses a wide range of products with vastly different engineering properties. Specifications for the structural design of wood products have been tied to elastic analysis and allowable stress design. Many of the behavioral equations used in ASD for members and connections are dated and in need of revision. Most schools of engineering do not teach wood design.

One of the most significant problems in using wood is that its strength is sensitive to the rate of load application and load duration, higher strengths being associated with higher rates and shorter durations of load. This problem is important enough that it must be accounted for in design. The "duration-of-load" (DOL) effect currently is taken into account by (1) basing allowable stresses on an assumed 10-year cumulative duration of live load; and (2) adjusting allowable stresses for other load combinations. The relation between duration of load and strength used to adjust stresses in the National Design Specification (NFPA, 1986), or NDS, is over 40 years old and is based on flexural tests of small clear specimens (Wood, 1951). In the 1970's, evidence began to accumulate that the DOL effect in small clears was not the same as that in dimension lumber and since then, several research programs have developed alternative approaches for dealing with DOL (Barrett and Foschi, 1979; Gerhards and Link, 1986). At the same time, modern stochastic load modeling (e.g., Turkstra and Madsen, 1980) has revealed that the 10-year assumed cumulative duration does not provide a supportable basis for assigning allowable stresses or strengths (Ellingwood, et al, 1988).

With a better understanding of the DOL effect, the availability of stochastic load models, and the results of an in-grade test program (IGTP) conducted in the early 1980's to determine the mechanical properties of nominal 2-inch dimension lumber (Green and Evans,1987), code preparation for engineered wood products could advance in the same direction as steel and concrete. Limit states design for wood was identified as a priority item at a workshop in 1983 (Itani and Faherty, 1984). An ASCE Task Committee was formed to develop a pre-standardization report on LRFD for wood; this report was completed in 1987 (Murphy, 1988). Finally in 1987, a consortium of companies in the wood industry of the United States and Canada initiated a research program to develop an LRFD specification for engineered wood construction. This program provided an opportunity to examine the basis of the current wood design provisions in the NDS, to create a design method that is consistent across product lines, and to develop new reliability-based design procedures. With the experience gained from the steel LRFD Specification, it was possible to complete a draft specification by June, 1990. This draft currently is in a trial design phase, and current plans are to process it as an ASCE consensus standard during the next year or two.

3. RELIABILITY ANALYSIS

The conceptual basis for probability-based design is founded in classical reliability theory (e.g., Freudenthal, et al, 1966; Shinozuka, 1983). Structural performance is described by a mathematical model relating the resistance and load variables for the specific limit state of interest. This model,

$$G\ (X_1,\ X_2,\ldots,X_m) = 0 \qquad (7)$$

in which $\mathbf{X} = (X_1,\ldots,X_m)$ = random resistance or load variables, is defined, by convention, such that failure occurs when $G(\mathbf{X}) < 0$. Thus, the limit state probability becomes,

$$P_f = \int \cdots \int f_X\ (x_1,x_2,\ldots,x_m)\ dx_1 dx_2 \ldots dx_m \qquad (8)$$

in which $f_X(\mathbf{x})$ = joint density function of **X** and the domain D is that region of **x** where $G(\mathbf{X}) < 0$.

Equation (8) is difficult to analyze for realistic structures. The occurrence of greater-than-design-basis events or structural failures is rare, and there seldom is sufficient data to define the distributions of **X** at the extreme values of significance for structural safety assessments. Moreover, the integral is difficult to evaluate. First-order reliability methods were developed in response to these problems (e.g., Hohenbichler, et al, 1987). In first-order reliability analysis, the reliability is measured by a reliability index, β, defined by,

$$\beta = m_G/\sigma_G \qquad (9)$$

in which m_G and σ_G are the mean and standard deviation of the limit state function, $G(\mathbf{X})$, linearized at an appropriate expansion point on the surface $G(\mathbf{X}) = 0$. The reliability index and limit state probability are related, approximately, as,

$$P_f = \Phi\ (-\beta) \qquad (10)$$

$$\beta = \Phi^{-1}(1 - P_f) \qquad (11)$$

in which $\Phi\ (\cdot)$ = standard normal probability integral. When the limit state is linear and the variables are normal, Eqs. (10) and (11) are exact.

As noted previously, the strength of wood is time-dependent and failure may occur by creep rupture as well as by overload. Thus, the entire load history must be known to perform the reliability analysis. This adds a degree of complexity to the reliability analysis of wood structures that is not encountered for steel and concrete structures, where the dependence of strength on time is negligible and only the probability distribution of the maximum load needs to be known (Galambos, et al, 1982; Ellingwood, et al, 1982). The creep rupture limit state is analyzed using static fatigue damage accumulation models. Several models have been developed for this purpose at the U.S. Forest Products Laboratory (USFPL) (Caulfield, 1985; Gerhards and Link, 1986) and at Forintek Canada (FC) (Barrett and Foschi, 1978; Foschi, et al, 1989). These models utilize a state variable, α (t), to measure damage. The state variable is normalized so that it increases (stochastically) with time from zero in the undamaged state to 1.0 at failure.

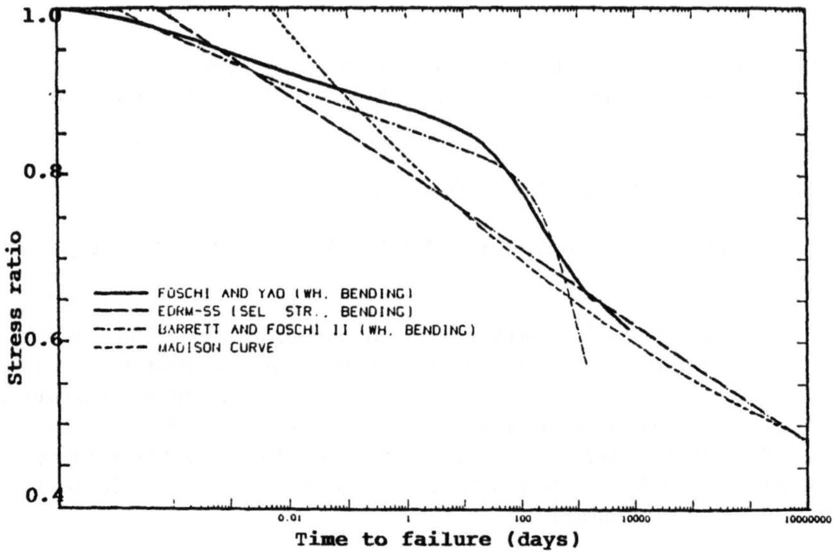

Figure 1. Comparison of Various Time-to-Failure Models under Constant Stress.

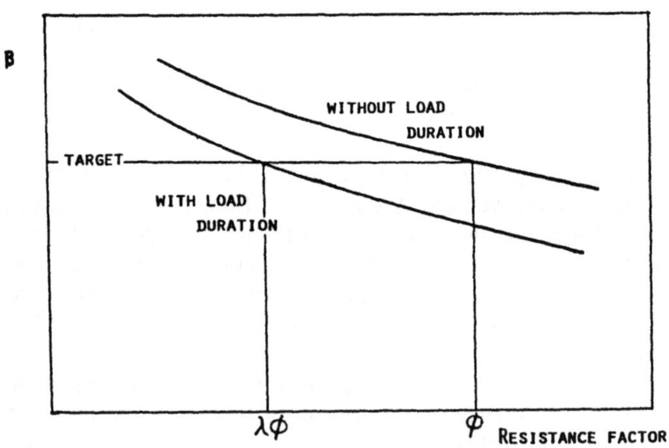

Figure 2. Determination of the Load Effect Factor for Maintaining Equal Reliability.

The USFPL model describes damage rate as an exponential function (Gerhards and Link, 1986),

$$d\alpha/dt = \exp\left[-A + B\,\sigma(t)\right] \qquad (12)$$

in which $\sigma(t)$ is the ratio of applied stress to short-term strength measured in a conventional strength test of approximately 5 to 10 minutes duration, and A and B are experimentally determined constants. The FC models are functions of the existing damage as well as stress (Barrett and Foschi, 1979; Foschi, et al, 1989);

$$d\alpha/dt = A\,(\sigma(t) - \sigma_o)^{B} + C\alpha \qquad (13)$$

$$d\alpha/dt = A\,(\sigma(t) - \sigma_o)^{B}\sigma^{B}_{s} + C\,\sigma^{n}_{s}\,(\sigma(t)-\sigma_o)^{n}\alpha \qquad (14)$$

in which parameters A, B, C and n are model constants, σ_s is the short-term strength, and σ_o is a damage threshold stress ratio below which no damage accumulates. Damage predicted by Eqs. (13) and (14) is similar above a stress ratio of about 0.6. As a point of interest, the DOL effect implied in the current NDS (Wood, 1951; NFPA, 1986) also can be expressed in terms of a damage rate:

$$d\alpha/dt = A\,(\sigma(t) - \sigma_o)^{B} \qquad (15)$$

The constants A, B, and C in Eqs. (11) - (14) are not the same, of course, and the models all have been calibrated to existing DOL test data. Times to failure under constant stress predicted by Eqs. (12) through (15) are compared in Figure 1. All DOL tests to date have been conducted using constant loads, which limits the applicability of the current DOL models to the analysis of load combinations involving static loads. The application of any of these DOL models to load combinations involving wind or earthquake loads is questionable, a fact not recognized in the current NDS (NFPA, 1986).

A stochastic model of the entire load process is required for analyzing damage accumulation using Eqs. (12) through (15). Simple Poisson square-wave and intermittent pulse process models (Turkstra and Madsen, 1980), in which the loads are modeled as sequences of random static loads, are sufficiently accurate to describe the variation in time of gravity loads due to the permanent weight of the structure and permanent attachments, occupancy live loads and snow loads (Ellingwood, et al, 1988). Typical statistical data on the maximum dead and live loads during reference periods of 1 year and 50 years are summarized in Table 1. The mean values are referenced to the nominal loads in ASCE Standard 7-88 (ASCE, 1990). Variations from site to site should be expected for environmental loads. For the analysis of stochastic damage accumulation, square-wave pulse process models of gravity loads are constructed to be consistent with the statistics in Table 1 (Ellingwood, et al, 1988; 1991).

Table 1. Statistical Data on Structural Loads

LOAD	1 YEAR			50 YEARS		
	Mean	cov	cdf	Mean	cov	cdf
Dead	1.05D	0.10	Normal	1.05D	0.10	Normal
Live	0.42L	0.45	Type I	1.00L	0.25	Type I

Table 2. Time Effect Factors, λ, for LRFD

Load Combination	USFPL	Forintek	Recommended
D (only)	0.59	0.50	0.60
D + L (light)	0.76	0.84	0.80
D + L (heavy)	0.67	-	0.70
D + L (roof)	0.78	-	0.80
D + S	0.75	0.83	0.80
D + W	>1.0	-	1.00
Impact	>1.00	-	1.00

The state variable, $\alpha(t)$, can be expressed as a summation of damage increments occurring during successive load pulses;

$$\alpha(t) = \sum_i \Delta\alpha_i \tag{16}$$

The limit state function for damage accumulation becomes,

$$G(\mathbf{X}) = 1 - \alpha(\mathbf{X},t) \tag{17}$$

The limit state probability is given as,

$$P_f = P[\alpha(t) > 1;\ 0 < t \leq T] \tag{18}$$

in which T is the period of reference, taken by convention to be 50 years in reliability studies in the United States (Galambos, et al, 1982; Ellingwood, et al, 1982). The probability of failure can be obtained by Monte Carlo simulation (Ellingwood, et al, 1988). The reliability index then is defined from Eq.(11).

4. RELIABILITY BASES FOR LIMIT STATES DESIGN

4.1 Assessment of Current Design Practice

The assessment of reliability measures associated with existing standards is necessary to guide the selection of target reliabilities upon which new criteria may be based. With a knowledge of the resistance and load statistics, the β or P_f associated with any given design can be computed, thereby establishing a series of reliability benchmarks. This process is known as code calibration. Criteria for limit states of tension or flexure involving gravity load combinations have been studied extensively. These limit states are common, gravity loads govern design in many practical cases, and there is general agreement that current design practice is satisfactory. The calibration was completed earlier for steel and reinforced concrete structures (Galambos, et al, 1982) in the process of developing the LRFD specification (AISC, 1986) and the uniform load requirements (ASCE 7-88). Comparable, if not identical, reliability measures should be expected for engineered wood construction.

The design strength (left hand side of Eq. (1) for a steel tension member, consistent with the target reliabilities identified above, is defined in LRFD by ASIC, 1986),

$$\text{Yield:} \quad 0.9\ F_y\ A_g \tag{19}$$

$$\text{Fracture:} \quad 0.75\ F_u\ A_e \tag{20}$$

in which F_y and F_u are yield and tensile strength, respectively, A_g is the gross area, and A_e is the effective net area. The reliability indices (on a 50-year basis) associated with these design strengths are about 2.5 for yielding and about 3.2 for fracture when $L_n/D_n = 2$, a typical value. Similarly, the design strength in flexure for a compact steel beam with full lateral support is,

$$\text{Flexure:}\quad 0.9\ F_y\ Z_x \tag{21}$$

in which Z_x is the plastic section modulus. For a typical L_n/D_n of 2, the associated β is approximately 2.5. A recent study to develop LRFD for cold-formed steel construction (Hsiao, et al, 1990) yielded β's in this general range as well.

The safety check for flexure in a reinforced concrete beam subjected to dead and live loads, designed by ACI Standard 318 (ACI, 1989), is

$$0.9\ M_{un} = 1.4\ D_n + 1.7\ L_n \tag{22}$$

in which M_{un} is the ultimate moment capacity computed using nominal material strengths and dimensions. Note that reinforced concrete has not yet adopted the common ASCE Standard 7-88 load requirements. The reliability index varies from 3.1 to 2.8 as L_n/D_n increases from 0.5 to 2.

In the current NDS (NFPA, 1986), wood beams are designed for dead plus live loads or dead plus snow loads using,

$$F_{b,\,all}\ S_x > D_n + L_n \tag{23}$$

$$1.15F_{b,\,all}\ S_x > D_n + S_n \tag{24}$$

in which $F_{b,\,all}$ designates the allowable stress in bending based on an assumed duration of 10 years for full design load and S_x is the elastic section modulus. The factor 1.15 accounts for the DOL effect for snow load, which is assumed to act cumulatively for 2 months rather than 10 years.

The stress ratio needed to analyze cumulative damage is equal to

$$\sigma(t) = [D + Q(t)]/[F_r S_x] \tag{25}$$

in which Q(t) is the stochastic live or snow load, as appropriate, and F_r is the short-term modulus of rupture, obtained by ramp-loading a member to failure over a period of approximately 5 to 10 minutes. For glulam beams in flexure, for example, F_r can be modeled by a two-parameter Weibull distribution, with mean and coefficient of variation of $2.86F_b$ and 0.15, respectively. This stress ratio is used with Eqs. (12) through (15) to compute reliabilities for a period of 50 years. For live load, $\beta = 2.7$, depending on the duration assumed for the transient live load (Ellingwood, et al, 1988). Similarly, for snow loads with pulse durations of about 1 week, and a probability of measurable snow on the roof during the snow season of 0.2, $\beta = 2.2$.

The current treatment of DOL in the NDS (NFPA, 1986) does not appear to be risk-consistent, with lower values of β resulting for snow loads. Bringing these two checks into alignment in LRFD will impact the relative cost of floors and roofs.

4.2 Target Reliability Measures

Reliability analyses of existing specifications indicate that values of β for members in flexure and/or compression tend to fall in the range of 2.4 to 3.0 (on a 50-year basis). The higher values are observed for gravity load combinations and the lower values are observed for combinations involving wind or earthquake load.

5. PROBABILITY-BASED LIMIT STATES DESIGN

A structural system or component can be defined by its nominal or characteristic strength properties, dimensions, and loads, expressed in vector form as $\mathbf{X}_n$. The limit state probability is a function of these variables, and so the code performance objective stated in probabilistic terms is,

$$\beta(\mathbf{X}_n) = \Phi^{-1}[1 - P_f(\mathbf{X}_n)] > \beta_o \tag{26}$$

in which β_o is the target reliability index. The goal of probability-based design is to arrive at an acceptable design, expressed by the set of design variables, $\mathbf{X}_n$, using conventional design equations of the form (see Eq.(1)),

$$\phi R_n > \sum_i \gamma_i Q_{ni} \tag{27}$$

in which R_n and Q_{ni} are nominal (code-specified) or characteristic resistance and load, respectively, which are functions of X_n, Φ is the resistance factor ($\gamma_m = 1/\phi$, the material factor), and γ_i are the load factors, which are independent of construction material. The transformation between Eqs. (26) and (27) is carried out by a code committee and the designer is not affected by it. It should be emphasized that the load and resistance factors in Eq. (27) account for inherent uncertainties in capacity and load under normal "error-free" conditions.

The right hand side of Eq. (27) has already been fixed by ASCE Standard 7-88 (see Eqs.(4) through (6)), so only the factors defining the design strength for an engineered wood product need be considered. The design strength must include the DOL effect. This is done by modifying the design strength to read:

$$\text{Design strength} = \gamma \, \phi \, R_n \tag{28}$$

The adjusted strength, R_n, is based on an agreed-upon percentile of the short-term strength in the end use condition, commonly about the 5-percentile value. The resistance factor, ϕ, takes into account inherent variability in short-term strength, as well as the mode and consequence of failure. The time effect factor, λ, accounts for

Designers are already familiar with most nominal values because they are used in both ASD and strength design. More important, most structural codes cross-reference other standards and codes for many of the nominal load and strength values. For example, when a designer uses the AISC LRFD specification, he also utilizes (by reference) the requirements of ASTM Standard A36 on hot-rolled steel. If the AISC Specification Committee were to alter the specification of the nominal yield stress to be used in design, either ASTM A36 would have to be altered as well, or it no longer could be referenced the Specification. Problems of a similar nature occur in all design specifications; ASTM D245 (1989) governing visually graded lumber properties is a case in point. The inter-relation of various codes and standards constrains the development of resistance values used in reliability-based design. Only a complete overhaul of the code and standard development process will solve this problem.

6. SOME CURRENT ISSUES

One argument in favor of probability-based design is that it provides a quantitative measure of safety. This has created a natural desire to validate probabilistic methods by comparing predicted to observed failure rates. There also is a temptation to make comparisons between the performance and safety of different materials, and to use these comparisons in marketing the product. Such comparisons seldom are valid at any but the most elemental levels of structural component complexity. Even at such levels, the differences between predicted and observed failure rates often are substantial. There are a number of reasons for this lack of agreement.

Structural safety studies invariably are concerned with rare events and the need to make inferences based on small sample statistics. Questions concerning the confidence in and interpretation of small probabilities have not abated during the past 20 years, even with the widespread use of first-order reliability methods developed, in large part, to circumvent them. Confidence intervals for the reliability estimates are quite large. Relative comparisons of reliability within a narrow domain of application, where the assumptions underlying the limit states can be tested for consistency, are more useful than comparisons across a wide domain, where the accuracy of behavioral assumptions may be vastly different from application to application.

The load combinations reflect only the effects of stochastic variability in loads. However, the majority of failures (reportedly 80 to 90 percent) are due to design and construction errors, which may not be amenable to conventional probabilistic modeling. Whether changing to LRFD will cause this failure rate due to error to decrease remains to be seen. While properly chosen safety factors may mask the effects of minor errors, few notable failures would have been prevented by changes to the load or resistance factors used in design. Quality assurance programs are a more appropriate way to address this problem.

the differences in the temporal characteristics of the loads and, in that sense, serves a similar purpose as the current allowable stress adjustments for DOL (NFPA, 1986). Suitable values of ϕ and λ can be obtained by designing members using Eq. (28), performing the reliability analysis with DOL included (failure due to creep rupture) and with DOL ignored (failure due to overload), and adjusting R_n, ϕ and γ until the desired β is obtained. This process is illustrated in Figure 2.

The design strength for four of the common limit states for engineered wood construction becomes:

Flexure:	$\lambda 0.85\ F_b S_x$	(29)
Compression:	$\lambda 0.90\ F_c A_g$	(30)
Tension:	$\lambda 0.80\ F_t A_n$	(31)
Shear:	$\lambda 0.75\ F_v A_{sh}$	(32)
Connections:	$\lambda 0.65\ Z_n$	(33)

in which F_b, F_c, F_t and F_v are nominal strengths in flexure, compression, tension and shear, S_x is the elastic section modulus, A_g is the gross area, A_n is the net area, A_{sh} is the shear area, and Z_n is the nominal connection strength. Table 2 summarizes the time effect factors obtained for the flexure limit state and different load combinations (Ellingwood and Rosowsky, 1991). Essentially the same DOL factors are obtained for these load combinations if the limit states of tension or compression parallel to grain rather than flexure are considered. The differences in the λ's are a reflection of the differences in the temporal characteristics of the stochastic loads. In a relative sense, the proposed factors for load combinations involving live and snow load in LRFD are different from the current ASD adjustments. Note that the USFPL and Forintek damage accumulation models result in different time effect factors.

Engineered wood construction involves a wide range of producers and products. Since the inherent variability of these products differs across a wide range, and manufacturing quality control varies from producer to producer, it is probable that the specified values of ϕ and R_n may have to vary somewhat across product lines in order to maintain the objective of a constant reliability index, β, and, and the same time, the relative market positions of various competing products. On the other hand, the time effect factor, λ, does not appear to be especially sensitive to variations in the load process parameters, to the strength statistics, or to small variations in the target reliability, β (Ellingwood and Rosowsky, 1991). Parameter λ depends primarily on the load combination considered and on the damage model selected.

The focus of attention in the above development has been on the partial factors of safety, λ and ϕ, rather than on the nominal values of the design variables.

Practical probability-based safety checks have focused so far on the behavior of individual members rather than on the structural system as a whole. As this is no different from what has been done traditionally, it is a reflection of the general state-of-the-art in codified design. However, one might expect that checks derived from a performance requirement for reliability of the structural system might differ appreciably from those developed so far. Conversely, failure rates of systems would differ from failure rates of individual isolated members.

System behavior is particularly important in light wood frame construction, where performance is affected by load sharing and partial composite action of members and sheathing. The NDS (NFPA, 1986) has recognized its importance for many years by permitting an increase of 15% in the allowable bending stress used in designing assemblies where three or more members are used repetitively. While this approach does not appear to have caused any particular problems, it lacks a sound rationale and thus cannot be extended to new applications or product lines. The analysis of reliability of structural systems requires that possible failure paths (or modes) be identified and an appropriate limit state be formulated for the system as a whole. Theoretical methods for analyzing reliability of structural systems have developed rapidly; practical insights and recommendations on implementation in design have followed more slowly.

Safety checking, by and large, will continue to be performed on a member or component basis for ordinary building construction. Designers are familiar with this approach and it can be codified easily. There is some precedent for adjusting the member safety check to account for system effects, albeit in an essentially subjective fashion, through the 15% increase in allowable stress mentioned above, through K-factors for systems behavior in seismic-resistant design (ASCE, 1990), and similar actions. The design strength (Eqs.(1) and (28)) for a member within an assembly would be,

$$\text{Design strength} = \psi \lambda \phi R_n \qquad (34)$$

in which ψ is the system effect factor. The adjustment should be made to the resistance criteria rather than to the load criteria because the features that distinguish system reliability from member reliability depend strongly on the material and type of construction.

Serviceability limit states are becoming increasingly important in flexible modern structures (Ad Hoc Committee, 1986). The current requirement to limit static deflections due to live load-to-span/360 or some similar limit has been in use for over 150 years and, in many cases, is insufficient to ensure acceptable performance, particularly when the cause of unserviceability is objectionable occupant-induced vibrations. In the current building code climate, serviceability issues are best dealt with through guidance material provided in a commentary rather than through prescriptive code provisions.

7. CONCLUDING REMARKS

The task to complete a draft probability-based limit states design criteria for engineered wood construction in the United States was completed in about two years. These criteria take the inherent variability in the strength and stiffness of wood products and time-dependent nature of their strength under stochastic loads into account in a consistent and relatively simple way. However, it would be wrong to assume that the main thrust in LRFD was in selecting the resistance factors. The process of specification development provided the opportunity to re-examine the overall technical basis for all major provisions. Accordingly, significant advances in the science of wood behavior and design are incorporated in the LRFD specification as well. Among these are the modern treatment given to instability of columns, beam-columns and long-span beams without lateral bracing (Zahn,1988), and the development of new connection design procedures based on yield theory (e.g., McLain and Carroll, 1990).

Why switch to LRFD? The technical basis of the provisions, many of which are clouded or otherwise untraceable in the U.S. NDS, now has been updated. Applying partial factors of safety to the load and strength terms makes the sources of design uncertainty more apparent and better accounts for variability. Practicing engineers who are uncomfortable with wood as a construction material may be convinced to change their view of it, now that there is a sound technical basis for the provisions. LRFD has narrowed the range of reliabilities in practice and, as a result, economies in design can be realized without sacrificing safety and performance. Finally, LRFD is easier to teach. This will facilitate the introduction of wood design into engineering curricula; in the long term, this will do more to enhance the use of wood in building construction than any other measure.

Acknowledgement

Support for this work from a grant administered by Engineering Data Management, Inc., on behalf of the U.S. wood industry is gratefully acknowledged.

References

Ad Hoc Committee on Serviceability (1986). "Structural serviceability: a critical appraisal and research needs." **J. Str. Engr., ASCE** 112(12):2646-2664.

American Concrete Institute (1989). "Building code requirements for reinforced concrete (ACI 318)." Detroit, MI.

American Institute of Steel Construction (1986). "Load and resistance factor design specification for steel buildings." Chicago, IL.

American Society of Civil Engineers (1990). "Minimum design loads for buildings and other structures (ASCE 7-88)." New York. (Previously Am. Nat.Std. A58.1-1982).

American Society for Testing and Materials (1989). "Standard practice for establishing structural grades and related allowable properties for visually graded lumber." ASTM D245, Philadelphia, PA.

Barrett. J.D. and Foschi, R.O.,(1978). "Duration of load and probability of failure in wood. Parts I and II" **Canadian J. of Civil Engr**., 5(4):505 - 532.

Caulfield, D.F. (1985). "A Chemical Kinetics Approach to the Duration-of-Load Problem in Wood." **Wood and Fiber Science** 17(4):504-521.

Ellingwood, B., MacGregor, J.G., Galambos, T.V. and Cornell, C.A. (1982). "Probability-based load criteria: load factors and load combinations." **J. Str. Div. ASCE** 108(5):978-997.

Ellingwood, B., Hendrickson, E.M. and Murphy, J.F. (1988). "Load Duration and Probability Based Design of Wood Structural Members." **Wood and Fiber Science** 20(2):250-265.

Ellingwood, B. and Rosowsky, D. (1991). "Duration of load effects in LRFD for wood construction." **J. Str. Engr., ASCE** 117(2):584-599.

Foschi,R.O., Folz, B. R. and Yao, F.Z. (1989). "Reliability based design of wood structures." Structural Research Series Report No. 34, Dept. of Civil Engr., Univ. of British Columbia, Vancouver.

Freudenthal, A., Garrelts, J. and Shinozuka, M. (1966). "The analysis of structural safety." **J. Str. Div. ASCE** 92(1):267-325.

Galambos, T.V., Ellingwood, B., MacGregor, J.G. and Cornell, C.A. (1982). "Probability-based load criteria: assessment of current design practice." **J. Str. Div. ASCE** 108(5):959-977.

Gerhards, C. and Link, C.(1986). "Effect of loading rate on bending strength of Douglas-fir 2 by 4's." **Forest Products Journal**, 36(2):63 - 66.

Green, D.W. and Evans, J.W. (1987). "Mechanical properties of visually graded lumber, Vols 1-5." USDA, Forest Service, Forest Products Laboratory, Madison, WI.

Hohenbichler, M., Gollwitzer, S., Kruse, W. and Rackwitz, R. (1987). "New light on first- and second-order reliability methods." **Str. Safety** 4(4):267-284.

Hsiao, L.E., Yu, W.-W. and Galambos, T.V. (1990). "AISI LRFD method for cold-formed steel structural members." **J. Str. Engr., ASCE** 116(2):500-517.

Itani, R. and Faherty, K., eds.(1984). "Structural wood research: state of the art and research needs." American Society of Civil Engineers, New York, 211 pp.

McLain, T.E. and Carroll, J.D. (1990). "Combined load capacity of threaded fastener wood connections." **J. Str. Engr., ASCE** 116(9):2419-2432.

Murphy, J.F., ed., (1988). "Load and resistance factor design for engineered wood construction - a prestandard report." Am. Soc. of Civil Engrs., New York, NY.

National Forest Products Association (1986). "National design specification for wood construction." Washington, DC.

Shinozuka, M. (1983). "Basic analysis of structural safety." **J. Str. Engr., ASCE** 109(3):721-740.

Turkstra, C. and Madsen, H. (1980). "Load combinations in codified structural design." **J. Str. Div., ASCE** 106(12):2527-2543.

Wood, L.W.(1951). "Relation of strength of wood to duration of load." Report R1916, Forest Products Laboratory, Madison, WI.

Zahn, J. (1988). "Combined load stability criterion for wood beam-columns." **J. Str. Engr., ASCE** 114(11):2612-2628.

FUNDAMENTALS OF RELIABILITY ASSESSMENTS

Henrik O. Madsen
Det norske Veritas, Danmark A/S,
Nyhavn 16,
DK-1051 Copenhagen K,
Denmark

ABSTRACT. This paper presents some fundamental reliability concepts and gives an introduction to commonly applied numerical methods for reliability evaluations. These methods are used not just to calculate failure probabilities, but also important sensitivity factors, updated failure probabilities when new information becomes available, and in connection with reliability based optimization. The presentation is concentrated on analysis of a single failure mode, where all uncertainties are described in terms of random variables. The methods can also be applied for system reliability evaluations and reliability evaluations involving random processes.

1. Introduction

In its simplest form a structural reliability problem is defined by a safety margin

$$M = R - S \tag{1.1}$$

formulated on basis of some mechanical model of the actual structure under analysis or design. The model includes a criterion of failure which allows for definition of a resistance R against failure to be compared with a maximum load effect S both related to some specified period of time. A fundamental assumption of structural reliability theory is that probability theory - in one or the other axiomatic set up - makes sense as the formal mathematical tool for modeling structural reliability concepts. This means that R and S may be thought of as having properties as random variables. In this case the structural reliability model should provide means to establish a joint probability distribution of (R, S). From this distribution the probability of the failure event $M<0$ or, alternatively, the probability of the survival event $M \geq 0$ may be calculated. A reasonable definition of the reliability of the structure for the given load situation is to let it be equal to the probability of survival.

For models of real highly reliable structures this theoretical failure probability is very small - of less order of magnitude as 10^{-3} per year, say. This means, that the failure probability is a very sensitive function of different choices of probability distributions of (R, S), all having the same mean value vector and covariance matrix. In a design situation with a given target failure probability of 10^{-5}, say, the consequence of this sensitivity is that quite varying structural dimensions occours, dependent on the choice of type of probability distribution. The engineer might ideally rely on available empirical data set to guide the choice. However, due to the small probabilities such data sets must be very large, in order to distinguish between different distribution tail

J. Bodig (ed.), Reliability-Based Design of Engineered Wood Structures, 21–45.

hypotheses with reasonable confidence. Real data sets are usually not much larger than to allow for an estimation of mean values, variances and covariances. Even this estimation possesses a considerable statistical uncertainty due to modest sample size, an uncertainty which should be included in the structural reliability analysis, of course. Perhaps the data allows for a confident choice between a limited number of distribution types, as what concerns the central part of the distribution. However, the data give no justification of using the tails of the selected mathematical distribution in the structural reliability analysis. Only occasionally a kind of mechanistic model is available from which the distribution type may be predicted. Examples of the most common mechanistic arguments are references to the central limit theorem and to the asymptotic theory of extremes. There are reasons to be very critical to such arguments, which often are too easily stated without proper justification of their basic premises. Asymptotic theories are laws of large numbers. In particular for the extreme value asymptotic theory, these large numbers must be extremely large in order that the asymptotic distribution are well approximated in the tail regions.

These differences are all summarized in the phrase "tail sensitivity problem". This problem causes the computed failure probability to be of limited informational value except for reliability comparisons made between the same model universe of probability distributions. Therefore, for the advancement of the use of modern probabilistic reliability analysis to aid structural engineering decisions in practice, there is an indispensable need for an agreement among competing engineers and the general public on using a standardized distribution model universe as a common reference. In other words, there is an indispensable need for a code of practice for structural reliability analysis.

An attempt to formulate such a code has recently been published by the Joint Committee on Structural Safety (JCSS) in the form of a working document with the title: Proposal for a Code for the Direct Use of Reliability Methods in Structural Design. JCSS is supported by the international associations CEB, CIB, ECCS, FIP, IABSE, IASS, and RILEM.

An important point in this code proposal is that the distribution types to be used in the reliability analysis are standardized. The way in which these standardizations are introduced is best illustrated by direct quotation from the proposal. In Chapter 7 on action modelling the following code type text is given:

> *Standardized distribution and process types to be used in action models for specific reliability investigations can be given in an action code to be used in parallel with this code on reliability methods. In such cases the action load model standardization given in this code are secondary to the standardizations of the action code.*

and similarly in section 8 on structural resistance modeling:

> *Standardized distributions of material properties to be used in structural resistance models can be given in material oriented codes to be used in parallel with this code on reliability methods. Standardized distributions given in such material codes are superior to the standardizations given in this code. It is required that a standardized distribution of a material property assigns zero probability to any set in which no value is possible due to the physical definition of the considered material property.*

In Section 9 on reliability models there is the following code type text:

> *If no specific distribution type is given as standard in the action and material codes this code for the purpose of reliability evaluations standardizes the clipped (or, alternatively, the zero-truncated) normal distribution type for basic load pulse amplitudes. Furthermore, the logarithmic normal distribution type is standardized for the basic strength variables.*

Deviations from specific geometrical measures of physical dimensions as length are standardized to have normal distributions if they act at the adverse state in the same way as load variables (increase of value implies decrease of reliability) and to have logarithmic normal distribution if they contribute to the adverse state in the same way as resistance variables (decrease of value implies decrease of reliability).

and further

In special situations other than the code standardized distribution types can be relevant for the reliability evaluation. Such code deviating assumptions must be well documented on the basis of a plausible model that by its elements generates the claimed probability distribution type. Asymptotic distributions generated from the model are allowed to be applied only if it can be shown that they by application on a suitable representative example structure lead to approximately the same generalized reliability indices as obtained by application of the exact distribution generated by the model.

Experimental verification without any other type of verification of a distributional assumption that deviates strongly from the standard is only sufficient if very large representative samples of data are available.

Distributional assumptions that deviate from those of the code must in any case be tested on a suitable representative example structure. By calibration against results obtained on the basis of the standardizations of the code it must be guaranteed that the real (the absolute) safety level is not changed significantly relative to the requirements of the code.

When arguing within a specific anticipatory model universe it is important to ensure that near zero probability value results are used for comparisons only within the model itself. Carrying the results to the outside world and attaching the usual probability interpretation of relative frequency of occurrence in the real world of the considered event, will generally be highly misleading even though the model has been carefully calibrated to real world data. This insight is not new. It has, however, now become urgent to focus on this point due to the recent maturing of practicable reliability analysis methods as described in the following sections.

This paper first gives some fundamental definitions followed by a description of the origin of various uncertainty types. The first- and second-order reliability methods as well as simulation methods are then described for calculation of failure probabilities and various sensitivity factors. Reliability updating as well as reliability based optimization is treated in some detail and the paper closes with a section on random process concepts in structural reliability. To ease the reading, references are not made in the text, but instead a list of the most important general text books and codes is given.

2. Some Fundamental Definitions

Physical formulation space: The n-dimensional space of points with coordinates that are the basic input variables to the considered structure and its environmental conditions (e.g. load variables, strength variables, geometrical variables). Time variation of the physical variables may often be taken into account on a level of idealization that corresponds to a discretisation to a finite set of time points. Thus each type of time varying basic variable contributes to the dimension of the formulation space with a number of basic input variables corresponding the number of time points. Generalization to continuous time is discussed in a separate section.

Set of possible observations: A given subset Ω of the physical formulation space. Points of the complement to this set are assumed to be excluded as possible observations of the vector of basic physical variables (e.g. negative compression strength etc.).

Transformed formulation space: The image by a given one-to-one normal continuous differentiable mapping T of the set of possible observations Ω into the n-dimensional normal space. The image $T(\Omega)$ is the entire space. It is assumed that Ω has sufficient regularity for such a mapping to exist.

Safe set: Subset S of Ω of points for which the structure is considered to suffer no failure. The same terminology is used for the image $T(S)$.

Failure set: Complement F of S relative to Ω. The same terminology is used for the image $T(F)$.

Failure surface (or limit state surface): The boundary between the safe set S and the failure set F. It is considered to be a subset of the failure set F. The same terminology is used for the boundary of $T(S)$.

Safety margin: A function M of the basic input variables with the property that $M>0$ for all points in the safe set, $M<0$ for all points in the failure set, and $M=0$ for all points on the failure surface.

Linear safety margin: A safety margin which is linear in all the uncertain input variables of the transformed formulation space.

Uncertain basic input variables of the physical formulation space are written jointly as the vector $\mathbf{Z} = (Z_1, \cdots, Z_n)$.

Uncertain basic input variables of the transformed formulation space are written jointly as the vector $\mathbf{U} = (U_1, \cdots, U_n)$. We have that $\mathbf{U} = T(\mathbf{Z})$ and $\mathbf{Z} = T^{-1}(\mathbf{U})$ where T is the mapping from subset Ω of the physical formulation space to the transformed formulation space.

Expectation (or mean value): Location parameter $E[Z]$ of uncertain quantity Z. All basic input variables of the transformed formulation space are assigned expectations:

$$\mathrm{E}[\mathbf{Z}] = (\mathrm{E}[Z_1], \ldots, \mathrm{E}[Z_n]) \tag{2.1}$$

The expectation is defined on all linear combinations of $Z_1, \cdots, Z_n$ as a linear functional:

$$\mathrm{E}[\mathbf{a}'\mathbf{Z}] = \mathbf{a}'\mathrm{E}[\mathbf{Z}] \tag{2.2}$$

in which the prime means the transposed and $\mathbf{a}$ is an arbitrary column matrix of constants.

Standard deviation: Dispersion parameter (or scale parameter) $\mathrm{D}[Z]$ of uncertain quantity Z.

Variance: $\mathrm{Var}[Z]$ = square of $\mathrm{D}[Z]$.

Covariance: Coupling parameter $\mathrm{Cov}[Z_1, Z_2]$ between two uncertain quantities Z_1 and Z_2. All pairs of basic input variables of the transformed formulation input space are assigned covariances $\mathrm{Cov}[Z_i, Z_j]$, $i,j = 1,\ldots,n$. The covariance is given properties like an inner product and it satisfies the Cauchy-Schwartz inequality

$$\mathrm{Cov}[Z_i, Z_j]^2 \le \mathrm{Cov}[Z_i, Z_i]\mathrm{Cov}[Z_j, Z_j] \tag{2.3}$$

The variances $\mathrm{Var}[Z_i]$ are defined as:

$$\mathrm{Var}[Z_i] = \mathrm{Cov}[Z_i, Z_i] \;,\; i=1,\ldots,n \tag{2.4}$$

According to the inner product properties, the covariance is defined between any two linear combinations of $Z_1, \cdots, Z_n$ by the formula:

$$\mathrm{Cov}[\sum_{i=1}^{n} a_i Z_i, \sum_{j=1}^{n} b_j Z_j] = \sum_{i=1}^{n}\sum_{j=1}^{n} a_i b_j \mathrm{Cov}[Z_i, Z_j] \qquad (2.5)$$

Any constant is assigned mean value equal to the constant itself and zero variance. Thus the covariance between a constant and any quantity is zero.

Correlation coefficient:

$$\rho[Z_i, Z_j] = \frac{\mathrm{Cov}[Z_i, Z_j]}{D[Z_i]D[Z_j]} \; ; \; D[Z_i]D[Z_j] \neq 0 \qquad (2.6)$$

$$\rho[Z_i, Z_j] = 0 \Leftarrow\Rightarrow \mathrm{Cov}[Z_i, Z_j] = 0$$

Coefficient of variation:

$$V_Z = \frac{D[Z]}{E[Z]} \; ; \; E[Z] > 0 \qquad (2.7)$$

Correlation matrix:

$$\rho_Z = \rho[Z_i, Z_j] \qquad (2.8)$$

Normal density function of mean vector $E[\mathbf{Z}]$ and regular covariance matrix $\mathbf{C_Z}$:

$$f_Z(\mathbf{z}) = \frac{1}{\sqrt{(2\pi)^n |\mathbf{C_Z}|}} \exp[-\frac{1}{2}(\mathbf{z}-E[\mathbf{Z}])'\mathbf{C_Z}^{-1}(\mathbf{z}-E[\mathbf{Z}])] \qquad (2.9)$$

in which $|\mathbf{C_Z}|$ is the determinant of $\mathbf{C_Z}$. For $E[\mathbf{Z}]$ = zero vector, and $\mathbf{C_Z} = \mathbf{I}$ = unit matrix, the density is rotational symmetric about the origin. It is called the n-dimensional standardized normal density:

$$f_Z(\mathbf{z}) = \frac{1}{\sqrt{(2\pi)^n}} \exp[-\frac{1}{2}(\mathbf{z}'\mathbf{z})] = \phi(z_1)\cdot \cdots \cdot\phi(z_n) \qquad (2.10)$$

Here $\phi(z)$ is the standard notation for the one-dimensional standardized normal density function defined by:

$$\phi(z) = \frac{1}{\sqrt{(2\pi)}} \exp(-\frac{1}{2}z^2) \qquad (2.11)$$

For the two-dimensional (bivariate) normal density of zero mean vector, unit standard deviations, and correlation coefficient ρ the notation $\phi(z,y;\rho)$ is used. It is:

$$\phi(z,y;\rho) = \frac{1}{2\pi\sqrt{(1-\rho)}} \exp[-\frac{1}{2}\frac{z^2-2\rho zy+y^2}{1-\rho^2}] \qquad (2.12)$$

Standardized normal distribution function $\Phi(z)$

$$\Phi(z) = \int_{-\infty}^{z} \phi(t)dt \qquad (2.13)$$

Distribution function for n-dimensional normal distribution with zero mean value vector, unit variances and correlation matrix ρ_Z.

$$\Phi(\mathbf{z}, \rho_Z) = \int_{-\infty}^{z_1} \cdots \int_{-\infty}^{z_n} \phi(\mathbf{t}, \rho_Z)dt_n \cdots dt_1 \qquad (2.14)$$

Conditional probability density function of Z_2 conditioned on Z_1

$$f_{Z_2|Z_1}(z_2|z_1) = \frac{f_{Z_1|Z_2}(z_1|z_2)}{f_{Z_2}(z_2)} \tag{2.15}$$

3. Uncertainty Sources

3.1 PHYSICAL UNCERTAINTY

The most obvious type of uncertainty, that affects the safety of structures, is the uncertainty of the material matter itself showing up as more or less random fluctuations of the physical properties from sample to sample. This type of uncertainty is called physical uncertainty. It may be measured in terms of relative frequencies of observing values of the physical characteristics in specified intervals or other relevant sets.

3.2 STATISTICAL UNCERTAINTY

Information on the physical uncertainty is obtained through a sample of observations of the relevant physical quantity. This information may be specified without essential uncertainty only if the sample size is large. Frequently only a small sample of observations is available. The uncertainty of the specifications of the measures of the physical uncertainty is called statistical uncertainty. Just as the physical uncertainty itself, the statistically uncertainty should be taken into account in the reliability analysis.

Within the framework of reliability theory the Bayesian statistical model for inference seems particularly attractive. Assume that information about a random variable Z, with known distribution type but unknown distribution parameters, is given through a sample $z_1, \cdots, z_k$ of k independent observations of Z. Let the mean value μ and the standard deviation σ be unknown. According to the Bayesian model, the mean value and standard deviation of Z are modeled as random variables M and Σ. Based on prior information about (M,Σ), a prior density $f_{M,\Sigma}(\mu,\sigma)$ is assigned. According to Bayes theorem, the posterior density is

$$f'_{M,\Sigma}(\mu,\sigma) = C\, f_{M,\Sigma}(\mu,\sigma) \prod_{i=1}^{k} f_z(z_i \mid \mu, \sigma) \tag{3.1}$$

where C is a normalizing constant. The product is called the likelihood function.

This Bayesian method dealing with statistical uncertainty requires a choice of a prior distributions of unknown parameters. Even though there are prior distributions which effectively may be considered to be noninformative, general professional knowledge may very well justify the choice of prior distribution that carries some information. Mathematical convenience points to the family of natural conjugate distributions associated with the distribution family of the input variables. Canonizing this distribution family, professional judgements concentrate on assessing location and scale parameters. This means that the prior distributions become no more than conceptional vehicles for transforming information based on professional judgement into reliability models.

3.3 MODEL UNCERTAINTY

Decisions based on structural reliability analysis depend, naturally, on the mathematical model which is set up for the analysis by the engineer. However, if careful real life decisions are made, it is necessary that considerations about the uncertainty of the model itself are quantified within the model. Model uncertainty can only be quantified either by comparisons with other more involved models that exhibit a closer representation of nature, or by comparisons with collected data from the field or laboratory. These so-called real data are, however, also representatives of model outputs, because behind any performance of data collection and data processing, there is some model which is never an error free and, much less, a complete model of reality. Consistent with this view, uncertainty due to less perfect measuring procedures is classified as model uncertainty.

Model selection is guided by a balance between ability to represent reality and the pragmatic need for having such simple mathematical properties of the model, that a large variety of problems can be analysed by the model. It is obvious from this that it is not particularly helpful to deal with the uncertainty of the simple model by actually calculating the differences of the results from the simple and the more complicated model. Model uncertainty should be introduced in such a way, that the pragmatic level of simplicity is not affected severely. Further, it is convenient if model uncertainty can be represented in a form where it is invariant to the mathematical transformations of the equations of the model. This is the case if it can be directly connected to the set of basic input variables of the model.

Model uncertainty in a structural reliability problem is due to two sources. First, the number of basic physical variables has been limited to a finite number n, leaving out possibly an infinite set of parameters that in the model idealization process have been judged to be of secondary or negligible importance for the problem at hand. By the realization of the structure in its environments of lifetime duration, the set of neglected parameters take on some set of values. For this set of values given, there is a subset of the formulation space, in which outcomes of the n explicitly considered variables will not cause failure. This is the safe set conditional on the outcome of the neglected variables. For another outcome of these variables, another slightly different safe set will result. Obviously the neglected variables act as generators of a background noise, the mechanism of which is usually only partly known. Thus the safe set may be modeled as a random set in accordance with a suitable probability law.

Second, model uncertainty is due to the idealization down to operational mathematical expressions. Besides this cause of pragmatic simplification, it may be due to lack of knowledge about the detailed interplay between the considered variables. For a given set of values of the neglected parameters, the lack of knowledge beyond the actual modeling of the limit state surface invites to consider the "true" failure surface as some perturbation of the idealized limit state surface. If this perturbation is considered to be an unknown element from a set of possible perturbations, an evaluation of the uncertainty may be given as some deviations from the idealized surface in terms of the entire ensemble of perturbations. In this view, also the second source of uncertainty can be modeled probabilistically, even though the adjoined probability measure should not be interpreted in the relative frequency sense.

Irrespective of the uncertainty source, the above discussion points out, that model uncertainty may be modeled as a deformation of the space by which the idealized limit state surface deforms randomly into "a possible true" limit state surface. Professional judgements are hardly fitted to point out specific distribution types for this purpose on the basis of objective evidence. Mathematical convenience is effectively the sole guidance for the choice. However, it may be possible to give judgmental assessments of location and scale parameters such as the mean and standard deviation. The basis for this is general engineering experience from working with relevant idealized models and from comparing results with observed data or other predictions calculated by use of more detailed models. Obviously such formal model uncertainty distributions play the same role as the prior distribution in the Bayesian statistical methods. Therefore judgemental random variables are occasionally called Bayesian random variables.

3.4 GROSS ERRORS

Uncertainty related to potential possibilities of occurrence of non-imaginable gross errors is of quite different nature than the three types of uncertainty discussed above. It seems not to be obvious how to include this uncertainty source in structural reliability analysis. Fortunately, there are reasonable arguments for keeping gross error analysis separated from the structural reliability analysis that takes account of physical uncertainty, statistical uncertainty and model uncertainty.

4. First And Second Order Reliability Analysis

4.1 SINGLE FAILURE MODE

In this chapter first- and second order reliability methods FORM/SORM are outlined for calculation of failure probabilities. All uncertain quantities are represented by random variables with an arbitrary distribution. The governing parameters (random and deterministic) are called the *basic variables* and they are denoted here by Z_i. The basic variables include loading parameters; strength parameters; and geometrical, statistical, and model uncertainty variables. For the application of FORM/SORM, the number of basic variables must be finite. Further, it must be possible for each set of values of the basic variables to state whether or not the structure has failed. This leads to a unique division of z-space into two sets, called the *safe set* S and the *failure set* F, respectively. The two sets are separated by the *failure surface* (or *limit state surface*), see Fig.1. The failure surface separating the safe set and the failure set is denoted by L_Z. A function $g(z_i)$ is called a *failure function* (or *limit state function*) if

$$\begin{aligned} g(z_i)&>0, \quad z_i \in S \\ g(z_i)&=0, \quad z_i \in L_Z \\ g(z_i)&<0, \quad z_i \in F \end{aligned} \tag{4.1}$$

According to this definition a failure function specifies the failure surface L_Z and satisfies a sign convention outside L_Z. The failure surface, on the other hand, does not define a unique failure function and care must be taken not to introduce a certain arbitrariness through the failure function. A simple choice for the failure function is

$$g(z_i) = \begin{cases} 1, & z_i \in S \\ 0, & z_i \in L_Z \\ -1, & z_i \in F \end{cases} \tag{4.2}$$

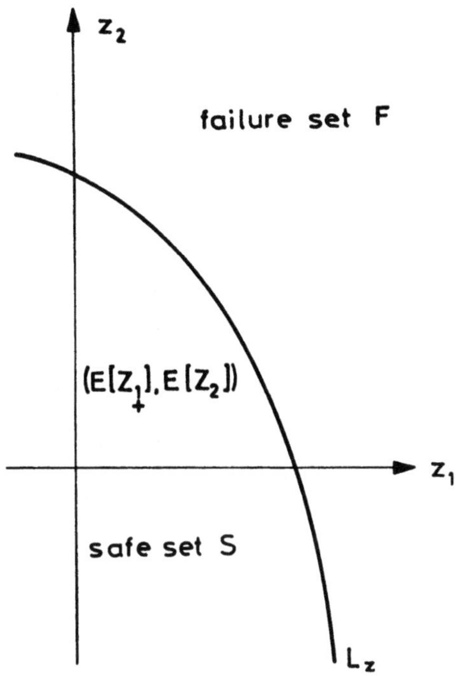

Fig 1 Geometrical illustration of the limit state surface

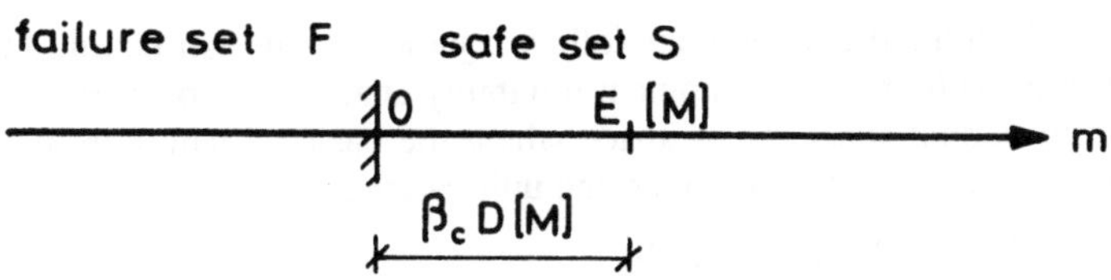

Fig 2 Geometrical illustration of the Cornell reliability index β_C.

For computational reasons a differentiable g-function is generally chosen whenever possible. The g-function usually results from the use of a mechanical analysis method for the structure.

The random variable obtained by replacing the parameters z_i in the failure function with the corresponding random variables Z_i is called a *safety margin* and is denoted by M:

$$M = g(Z_i) \tag{4.3}$$

This safety margin, by definition, reflects the arbitrariness introduced by the choice of a failure function.

It may be appropriate here to shortly describe the reliability index proposed by C. Allin Cornell who promoted the basic idea of second moment reliability philosophy. This philosophy suggests that all types of uncertainties entering the structural reliability problem should be expressed solely in terms of location (first moments) and scaling (second moments), which all should be combined into a single numerical measure of reliability, the reliability index. Besides escaping the tail sensitivity problem, a considerable operational simplicity is obtained by carring through the idea of basing reliability theory on the uncertainty algebra (which is a suitable name for the set of rules for operating with first and second moments). Based on the mean value and standard deviation of the safety margin C. Allin Cornell defined a *second moment reliability index* (or *safety index*) β_C as

$$\beta_C = \frac{\mathrm{E}[M]}{\mathrm{D}[M]} \tag{4.4}$$

This definition is illustrated geometrically in Fig. 2. In this one-dimensional case the failure surface is simply the point m=0. The idea behind the reliability index definition is that the distance from the location measure $\mathrm{E}[M]$ to the limit state surface provides a good measure of reliability. The distance is measured in units of the uncertainty scale parameter $\mathrm{D}[M]$.

The definition by C. Allin Cornell works well when the safety margin is a linear function of the basic variables. For safety margins which are non-linear functions of the basic variables, the definition of C. Allin Cornell had to be extended. Therefore, the second moment reliability indices of Hasofer and Lind and the generalized index of O. Ditlevsen were introduced. Many of the ideas developed in connection with these second moment indices have been carried over to the reliability index to be defined in this section for safety margins with basic variables of arbitrary distribution type.

When the basic variables follow a joint normal distribution, but the safety margin is non-linear, ideas from second moment reliability analysis can be used. A transformation into a standardized normal space, i.e. a space where the basic variables are normally distributed, mutually independent and with zero mean and unit variance is

$$\mathbf{U} = T(\mathbf{Z}) = \mathbf{A}(\mathbf{Z} - \mathrm{E}[\mathbf{Z}]) \tag{4.5}$$

where the matrix $\mathbf{A}$ obeys

$$\mathbf{I} = \mathbf{A}\mathbf{C}_{\mathbf{Z}}\mathbf{A}^T \tag{4.6}$$

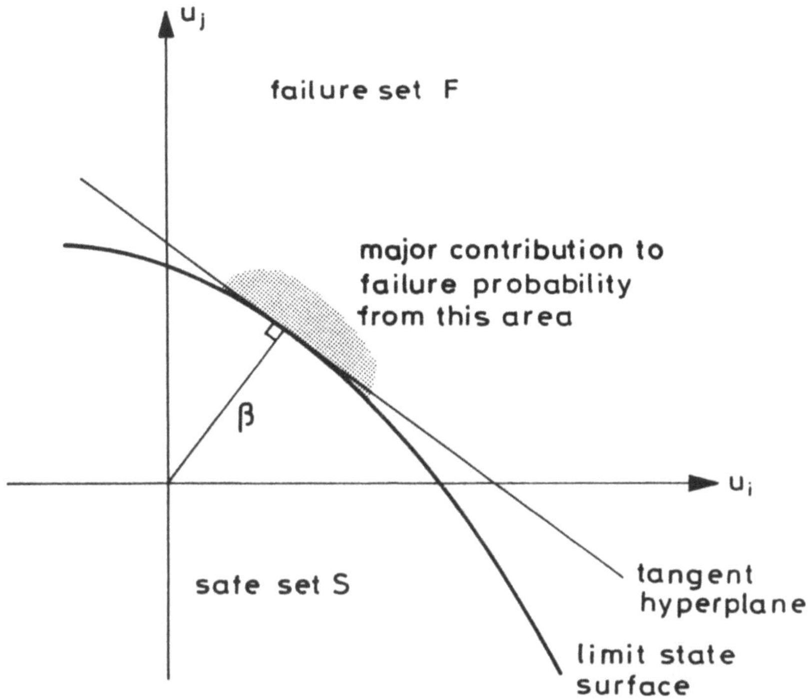

Fig 3 Illustration of approximate failure probability calculation

The probability density function of the vector **U** is

$$f(u_1,u_2,\cdots,u_n) = \phi(u_1)\phi(u_2)\cdots\phi(u_n);\quad \mathbf{u}\in R^n \tag{4.7}$$

where $\phi(u)$ is the probability density function for a normally distributed random variable with zero mean and unit variance.

$$\phi(u) = \frac{1}{\sqrt{2\pi}}\exp(-\frac{1}{2}u^2) \tag{4.8}$$

Any orthogonal transformation of the U-set results in a new set of normalized and uncorrelated variables. The distribution of the U-set is thus rotationally symmetric with respect to the second-moment distribution.

By the same mapping,

$$\mathbf{u} = \mathbf{A}(\mathbf{z}-\mathrm{E}[\mathbf{Z}]) \tag{4.9}$$

the mean-value point in z-space is mapped into the origin of u-space, and the failure surface L_Z in z-space is mapped onto the corresponding failure surface L_U in u-space as shown in Fig. 3. Due to the rotational symmetry of the second-moment representation of the U-set, it therefore follows that the geometrical distance from the origin in u-space to any point on L_U is simply the number of standard deviations from the mean-value point in z-space to the corresponding point on L_Z.

As illustrated in Fig 3, an approximation of the limit state function by its tangent hyperplane at the 'design point' u^* in the surface closest to the origin leads to a good approximation to the failure probability. This approximation is

$$P_F \approx \Phi(-\beta) \tag{4.10}$$

where $\beta = |u^*|$ is the smallest distance from the origin to the limit state surface. A better approximation results in replacing the approximating hyperplane by a second order surface, with the same tangent hyperplane and curvatures. This approximation to P_F is

$$P_F \approx \phi(\beta)\,\mathrm{Re}\left\{ i\left(\frac{2}{\pi}\right)^{1/2}\int_0^{i\infty} u^{-1}\exp\left(\frac{(u+\beta)^2}{2}\right)\prod_{j=1}^{u-1}(1-\kappa_j u)^{-1/2}du \right\} \tag{4.11}$$

in which $i=\sqrt{-1}$ and κ_j are the principal curvatures. A simple asymptotic approximation to (4.11) is

$$P_F = \Phi(-\beta) \prod_{i=1}^{n-1} (1-\beta\kappa_j)^{-1/2} \tag{4.12}$$

A reliability index β is defined as

$$\beta = -\Phi^{-1}(P_F) \tag{4.13}$$

For the tangent hyperplane approximation to the limit state surface, this reliability index is equal to $|u^*|$.

In general the basic variables are not normally distributed. The fact that probability contents in various sets are well approximated in a standardized normal space leads to the idea of finding a one-to-one transformation

$$\mathbf{T}: \mathbf{Z} = (Z_1, Z_2, \ldots, Z_n) \rightarrow \mathbf{U} = (U_1, U_2, \ldots, U_n) \tag{4.14}$$

where the random variables $U_1, U_2, \ldots, U_n$ are uncorrelated and standardized normally distributed. The limit state surface in z-space is mapped on the corresponding limit state surface in u-space. To evaluate the probability content in the failure set in u-space, search is performed for the minimum distance β from the origin to a point on the failure surface. The failure probability is then approximated to a first-order by $\Phi(-\beta)$, corresponding to a linearization of the failure surface. The linearization point is the design point $\mathbf{u}^*$. A reliability method based on this procedure is called a *first-order reliability method (FORM)*, and β is the *first-order reliability index*. Better approximation can be obtained by improved approximation of the failure surface, e.g., by a quadratic surface or a set of hyperplanes. The probability content in the approximating failure set can be evaluated by (4.11). A reliability method that uses a quadratic approximation to the failure surface at the design point is called a *second-order reliability method (SORM)*.

The simplest definition of the transformation T appears when the basic variables are mutually independent with distribution functions $F_{Z_1}, F_{Z_2}, \ldots, F_{Z_n}$. Each variable can then be transformed separately with the transformation defined by the identities

$$\Phi(u_i) = F_{Z_i}(z_i), \quad i=1,\ldots,n \tag{4.15}$$

The transformation is thus

$$\mathbf{T}: u_i = \Phi^{-1}(F_{Z_i}(z_i)), \quad i=1,\ldots,n \tag{4.16}$$

with the inverse transformation

$$\mathbf{T}^{-1}: z_i = F_{Z_i}^{-1}(\Phi(u_i)), \quad i=1,\ldots,n \tag{4.17}$$

The failure function g_u in u-space is given in terms of the failure function g in z-space as

$$g(\mathbf{z}) = g(\mathbf{T}^{-1}(\mathbf{u})) = g_u(\mathbf{u}) \tag{4.18}$$

The design point $\mathbf{u}^*$ is the solution to a minimization problem with one constraint:

$$\min_{g_u(\mathbf{u})=0} |\mathbf{u}| \tag{4.19}$$

Alternatively, $\mathbf{u}^*$ is determined as $\mathbf{u}^*=\mathbf{T}(\mathbf{z}^*)$, where $\mathbf{z}^*$ is the solution to

$$\min_{g(\mathbf{z})=0} |\mathbf{T}(\mathbf{z})| \tag{4.20}$$

Any algorithm solving this optimization problem can be used. In general, gradient based methods perform best. The value of the g_u-function is given in (4.18) and the partial derivatives are

$$\frac{\partial g_u}{\partial u_i} = \frac{\partial g}{\partial z_i}\frac{\partial z_i}{\partial u_i} = \frac{\partial g}{\partial z_i}(\mathbf{z})\frac{\phi(u_i)}{f_{Z_i}(z_i)} \tag{4.21}$$

$$= \frac{\partial g}{\partial z_i}(\mathbf{z})\frac{\phi(\Phi^{-1}(F_{Z_i}(z_i)))}{f_{Z_i}(z_i)}$$

where u_i and z_i are related through (4.15). Using a gradient based method, the failure function therefore needs not be expressed explicitly in terms of the u-variables, but all calculations are based on the g-function in terms of the original variables z_i. After each step in the algorithm, a new approximation to $\mathbf{u}^*$ is computed and the corresponding point $\mathbf{z}=\mathbf{T}^{-1}(\mathbf{u})$ is determined before the next step in the algorithm. The inverse transformation $\mathbf{T}^{-1}$ is often only given numerically, and this step can thus cause practical problems.

Let the point on the failure surface closest to the origin have the coordinates u_i^*. The tangent hyperplane to the failure surface at this point has the equation

$$\sum_{i=1}^{n} \frac{\partial g_u}{\partial u_i}(\mathbf{u}^*)(u_i-u_i^*) = 0 \tag{4.22}$$

and the first-order approximation to the failure probability P_F is

$$P_F \approx \Phi(-\beta) = \Phi(-|\mathbf{u}^*|) \tag{4.23}$$

Normalizing (4.22) leads to

$$\boldsymbol{\alpha}^T\mathbf{U}-\beta = 0 \tag{4.24}$$

and the $\boldsymbol{\alpha}$-vector is the unit normal vector to the tangent hyperplane

$$\boldsymbol{\alpha} = -\frac{\nabla g}{|\nabla g|} \tag{4.25}$$

where $\nabla g = \{\frac{\partial g}{\partial u_i}\}$ is the gradient vector.

The transformation (4.16) is only one out of an infinite set of possibilities, and the approximation in (4.23) depends on the choice of transformation. Practical experience indicates that the transformation in (4.16) is a good choice.

When the basic variables are not mutually independent, the *Rosenblatt transformation* has been suggested as a good choice. The transformation is defined similar to (4.16) as

$$\mathbf{T}: \quad \begin{aligned} u_1 &= \Phi^{-1}(F_1(z_1)) \\ u_2 &= \Phi^{-1}(F_2(z_2|z_1)) \\ &\cdot \\ u_i &= \Phi^{-1}(F_i(z_i|z_1,z_2,\ldots,z_{i-1})) \\ &\cdot \\ u_n &= \Phi^{-1}(F_n(z_n|z_1,z_2,\ldots,z_{n-1})) \end{aligned} \tag{4.26}$$

where $F_i(z_i|z_1,\ldots,z_{i-1})$ is the distribution function of Z_i conditional upon $(Z_1=z_1,\ldots,Z_{i-1}=z_{i-1})$:

$$F_i(z_i|z_1,\ldots,z_{i-1}) = \frac{\int_{-\infty}^{z_i} f_{Z_1,\ldots,Z_{i-1},Z_i}(z_1,\ldots,z_{i-1},t)dt}{f_{Z_1,\ldots,Z_{i-1}}(z_1,\ldots,z_{i-1})} \tag{4.27}$$

The transformation therefore first transforms Z_1 into a standardized normal variable. Then all conditional variables of $Z_2|Z_1=z_1$ are transformed into a standardized normal variable, and so forth. The Rosenblatt transformation is identical to the transformation in (4.16) when the basic variables are mutually independent. It can also be shown that the transformation is linear and of the form (4.10) when the basic variables are jointly normal.

The inverse transformation can be obtained in a stepwise manner as

$$\mathbf{T}^{-1}: \quad \begin{aligned} z_1 &= F_1^{-1}(\Phi(u_1)) \\ z_2 &= F_2^{-1}(\Phi(u_2)|z_1) \\ &\cdot \\ z_n &= F_n^{-1}(\Phi(u_n)|z_1,\ldots,z_{n-1}) \end{aligned} \tag{4.28}$$

In practical situations the transformations **T** and its inverse $\mathbf{T}^{-1}$ must often be determined numerically.

In u-space the failure function is given as in (4.18) and the partial derivatives are

$$\frac{\partial g}{\partial u_i}(\mathbf{u}) = \sum_{j=1}^{n} \frac{\partial g}{\partial z_j} J_{ji} \tag{4.29}$$

$\mathbf{J}=\{J_{ij}\}$ is the Jacobian matrix. The inverse matrix is given by

$$J_{ij}^{-1} = \frac{\partial u_i}{\partial z_j} = \begin{cases} 0, & i<j \\ \dfrac{f_i(z_i\emptyset z_1,\ldots,z_{i-1})}{\phi(u_i)}, & i=j \\ \dfrac{\frac{\partial F_i}{\partial z_j}(z_i\emptyset z_1,\ldots,z_{i-1})}{\phi(u_i)}, & i>j \end{cases} \tag{4.30}$$

u_i can be inserted from (4.26) and the Jacobian is then given in terms of $\mathbf{z}$. $\mathbf{J}$ and $\mathbf{J}^{-1}$ are lower-triangular matrices.

Two simulation methods called 'directional simulation' and 'importance sampling' are alternative methods to the analytical FORM/SORM. These simulation methods are also based on a transformation to a standardized normal space. The methods are more time consuming then FORM/SORM but provide an important tool for checking of FORM/SORM accuracy.

4.2 SERIES AND PARALLEL SYSTEM

For a series system of k components the failure probability is

$$P_F = P(M_1 \leq 0 \cup M_2 \leq 0 \cup \cdots \cup M_k \leq 0) \tag{4.31}$$

where $M_i = g_i(\mathbf{Z})$ is the safety margin for the ith component failure mode. In a first order approximation

$$P_F = 1 - \Phi(\boldsymbol{\beta}, \boldsymbol{\rho}) \tag{4.32}$$

where $\boldsymbol{\beta}$ is the vector of FORM reliability indices for the k safety margins, and the correlation coefficients is the matrix $\boldsymbol{\rho}$ are obtained from the individual $\boldsymbol{\alpha}$-vectors.

$$\rho_{ij} = \boldsymbol{\alpha}_i' \boldsymbol{\alpha}_j \tag{4.33}$$

For a parallel system of k component the failure probability is

$$P_F = P(M_1 \leq 0 \cap M_2 \leq 0 \cap \cdots \cap M_k \leq 0) \tag{4.34}$$

In a first order approximation

$$P_F = \Phi(-\boldsymbol{\beta}, \boldsymbol{\rho}) \tag{4.35}$$

where $\boldsymbol{\beta}$ is the vector of FORM reliability indices for the safety margins linearized at the joint design point, and $\boldsymbol{\rho}$ is as in (4.33) with the $\boldsymbol{\alpha}$-vector determined at the joint design point.

5. Sensitivity Analysis

Besides the reliability measure it is in many cases equally useful to know the value of the sensitivity factors, which in this section are described as related to a first-order reliability analysis for a single failure mode.

In the transformed standard normal space the approximating linear safety margin is

$$M = \beta - \boldsymbol{\alpha}^T \mathbf{U} \tag{5.1}$$

The variance of the safety margin is

$$Var[M] = \alpha_1^2 + \cdots + \alpha_n^2 = 1 \tag{5.2}$$

It is seen that α_i^2 is the fraction of the variance of the safety margin, which is caused by the ith standardized normal variable U_i. If the ith basic variable Z_i is independent of the other basic variables, α_i^2 is the fraction of the variance of the safety margin caused by the uncertainty in Z_i. If the basic variables are not independent, one must be more cautious in the interpretation of α_i^2 as this quantity is linked to several basic variables.

Another useful measure for the importance of the uncertainty in a basic variable is the *omission sensitivity factor.* This factor expresses the relative error in the first-order reliability index if a basic variable is replaced by a fixed value. If the ith basic variable is replaced by its median value the reliability index is increased by a factor called the omission sensitivity factor

$$\zeta = \frac{1}{\sqrt{1-\alpha_i^2}} \tag{5.3}$$

If another value than the median value is used, a slightly more complicated result applies. Two important practical applications of the omission sensitivity factor can be mentioned. Firstly, if the numerical value of α_i is less than 0.14, the relative error on the reliability index in replacing this variable by its median is less than 1%. In practical situations it is therefore possible to identify random basic variables, which can be replaced by a fixed value without introducing a significant error in the reliability index. Secondly, in the optimization routines applied to determine the design point, gradient based methods are generally used. The gradient must often be calculated by numerical differentiation and the computation time can be critical if the limit state function is complicated and the number of basic variables is large. From the general expression for the omission sensitivity factor it can be seen that if the transformed normal variable U_i is replaced by $\beta\alpha_i/2$, the relative error in the reliability index is of the order $o(\alpha_i^4)$, which is without practical importance if α_i is small. In the first iteration a fairly good estimate of α_i is often obtained, and in the following iterations variables with small α-values can be replaced by $\beta^{(m)}\alpha_i^{(1)}/2$, where the upper index signifies the iteration number. When the iteration has been completed, a control can be performed applying the full set of basic variables in a last iteration.

The perhaps most important sensitivity factor gives the sensitivity of the failure probability or the reliability index to variations of parameters in the safety problem. The parameters, denoted by $\mathbf{p}$, may include parameters in the distributions of the basic variables and deterministic parameters in the limit state function.

The derivative of the first-order reliability index with respect to a distribution parameter or a limit state function parameter is given here.

For a *distribution parameter* the result is

$$\frac{\partial}{\partial p_i}\beta(\mathbf{p}) = \frac{1}{\beta}\mathbf{u}^{*T}\frac{\partial}{\partial p_i}\mathbf{T}(\mathbf{z}^*,\mathbf{p}) \tag{5.4}$$

where $\mathbf{z}^*=\mathbf{T}^{-1}(\mathbf{u}^*,\mathbf{p})$.

For a *limit state function parameter* the result is

$$\frac{\partial}{\partial p_i}\beta(\mathbf{p}) = \frac{\frac{\partial}{\partial p_i}g(\mathbf{u}^*,\mathbf{p}_0)}{|\nabla g(\mathbf{u}^*,\mathbf{p}_0)|} \tag{5.5}$$

Let as an example the basic variables be mutually independent and let the basic variable Z_1 be normally distributed with mean value μ_1 and standard deviation σ_1. Z_1 and U_1 are related by

$$U_1 = T_1(Z_1) = \frac{Z_1-\mu_1}{\sigma_1} \tag{5.6}$$

The sensitivity factors with respect to μ_1 and σ_1 follow from (5.4) as

$$\frac{\partial\beta}{\partial\mu_1} = \frac{1}{\beta} u_1^* \frac{\partial}{\partial\mu_1}\left(\frac{z_1^*-\mu_1}{\sigma_1}\right) = -\frac{u_1^*}{\beta\sigma_1} \tag{5.7}$$

$$\frac{\partial\beta}{\partial\sigma_1} = \frac{1}{\beta} u_1^* \frac{\partial}{\partial\sigma_1}\left(\frac{z_1^*-\mu_1}{\sigma_1}\right) = -\frac{(u_1^*)^2}{\beta\sigma_1} \tag{5.8}$$

Let the basic variable Z_2 have a lognormal distribution with mean value μ_2 and coefficient of variation V_2. Z_2 and U_2 are related by

$$U_2 = T_2(Z_2) = \frac{\log Z_2 - \mathrm{E}[\log Z_2]}{\mathrm{D}[\log Z_2]} \tag{5.9}$$

where the mean value and standard deviation of $\log Z_2$ are

$$\mathrm{E}[\log Z_2] = \log\mu_2 - \frac{1}{2}\log(1+V_2^2) \tag{5.10}$$

$$\mathrm{D}[\log Z_2] = \sqrt{\log(1+V_2^2)} \tag{5.11}$$

The sensitivity factors with respect to μ_2 and V_2 follow from (5.4) as

$$\frac{\partial\beta}{\partial\mu_2} = -\frac{u_2^*}{\beta\mu_2\sqrt{\log(1+V_2^2)}} \tag{5.12}$$

$$\frac{\partial\beta}{\partial V_2} = -u_2^* \frac{V_2}{\beta\sqrt{\log(1+V_2^2)}(1+V_2^2)}\left(\frac{u_2^*}{\sqrt{\log(1+V_2^2)}} - 1\right) \tag{5.13}$$

The sensitivity factors can be extended to series and parallel systems. The simulation methods mentioned in Section 5 can also be extended to cover sensitivity factors.

6. Reliability Updating

Sensitivity factors for parallel systems are useful in several connections of which one is related to reliability updating.

During production of a structure or during its service life, additional information about the loading or strength often becomes available. This information is often of uncertain nature but can still be used to update the reliability. The additional information can be for a single basic variable in which case the distribution of this variable can be updated, or it can be in the form of a relation between several basic variables. In the latter case it can be expressed in terms of an event margin $h(\mathbf{Z})$ in one of two forms:

$$h(\mathbf{Z}) \leq 0$$

$$h(\mathbf{Z}) = 0 \tag{6.1}$$

where $\mathbf{Z}$ is the basic variable vector, possibly expanded to include variables related to e.g.

measuring uncertainty. With additional information of the first form, the updated failure probability P_F' is for a component with safety margin $g(\mathbf{z})$ is

$$P_F' = P(g(\mathbf{Z})\leq 0 \mid h(\mathbf{Z})\leq 0) = \frac{P(g(\mathbf{Z})\leq 0 \cap h(\mathbf{Z})\leq 0)}{P(h(\mathbf{Z})\leq 0)} \tag{6.2}$$

The numerator is thus calculated as the failure probability for a parallel system and the denominator as the failure probability for a component. With additional information of the second form the updated failure probability is

$$P_F' = P(g(\mathbf{Z})\leq 0 \mid h(\mathbf{Z})=0) = \frac{\frac{\partial}{\partial x}P(g(\mathbf{Z})\leq 0 \cap h(\mathbf{Z})\leq x)}{\frac{\partial}{\partial x}P(h(\mathbf{Z})\leq x)}\Bigg|_{x=0} \tag{6.3}$$

The numerator is thus calculated as the sensitivity factor for a parallel system and the denominator as the sensitivity factor for a component. Reliability updating performed in this way has proven useful in relation to e.g. inspection, proof loading and system monitoring and identification.

7. Reliability Based Optimal Design

Methods for structural optimization and methods for structural reliability analysis have in recent years developed considerably. Less work has been carried out to combine the two topics and thereby arrive at procedures for reliability based optimal design. It is therefore of considerable interest to combine the topics and to demonstrate the practical applicability and advantages for an actual design. The main application of such a methodology is expected to be for unique, expensive structures and for structures which are to be produced in large numbers.

Traditional structural optimization based on deterministic analysis is typically formulated as

min. $C(\mathbf{p},\mathbf{q})$

s.t. simple constraints on design parameters (7.1)

s.t. partial coefficient code requirements

The optimization aims at minimizing the objective function $C(\,)$ which is a function of the cost parameters $\mathbf{p}$ and the design parameters (optimization variables) $\mathbf{q}$. The objective function is often the cost of design and construction, while in recent years attempts to also include the cost of maintenance have been made. The optimization must obey constraints. Simple constraints on the design parameters and constraints from the code requirements are typically formulated. The code requirements can refer to both ultimate and serviceability limit states. This formulation does not explicitly include reliability requirements, but it is implicitly assumed that the reliability level obtained by use of the present codes is optimal. It is sometimes argued that otherwise these codes would have been changed.

In connection with a code for direct use of probabilistic methods, an optimization problem is often formulated as

min. $C(\mathbf{p},\mathbf{q})$

s.t. simple constraints on design parameters (7.2)

s.t. $P_F \leq P_F^{\max}$

The constraints related to code requirements in (7.1) are then replaced by a constraint on the failure probability P_F. The code defines this limit value for the failure probability and specifies the basis for computing it through specification of distribution types, inclusion of model uncertainty, inclusion of statistical uncertainty, load and load combination modeling, resistance modeling and system reliability modeling.

In a more advanced setting the optimization is formulated as

$$\begin{aligned} &min.\ C_T(\mathbf{p},\mathbf{q}) \\ &s.t.\ simple\ constraints\ on\ design\ parameters \end{aligned} \tag{7.3}$$

The objective function C_T is the total expected cost including the expected cost of failure. It is expressed as

$$C_T(\mathbf{p},\mathbf{q}) = C_I(\mathbf{p},\mathbf{q}) + C_F(\mathbf{p},\mathbf{q})P_F(\mathbf{q}) \tag{7.4}$$

$C_I(\,)$ is the cost of design, construction and expected cost of maintenance, while C_F is the cost of failure. The cost of failure should include both tangible and intangible cost.

It is recognized that the optimization aims at minimizing the expected total cost. It is therefore only the expected value of the cost items that enters the analysis. These costs are in practice only assessed with uncertainty and it is recommended to perform a sensitivity analysis to determine the change in the solution of the optimization to a change in the cost input parameters.

In this description the reliability index and failure probability are computed by first-order reliability methods and only one failure mode is considered. The limit state function for this mode is denoted by $g(\,)$ and is a function of the deterministic design parameters $\mathbf{q}$ and uncertain basic variables $\mathbf{u}$ (in the transformed standard normal space). The first-order reliability method determines the reliability index by solving an optimization problem

$$\begin{aligned} &min.\ |\mathbf{u}| \\ &s.t.\ g(\mathbf{u},\mathbf{q})=0 \end{aligned} \tag{7.5}$$

The solution is

$$\beta = |\mathbf{u}| \tag{7.6}$$

It is recognized that the optimization in (7.3) when applied in connection with a first-order reliability method has an objective function with a value determined by solving a second optimization.

The optimization can be formulated as a nested optimization, and an implementation is being done within the reliability analysis program PROBAN at Veritas Research. The advantages of formulating the analysis as a nested optimization are:

- an existing reliability analysis program can be used
- the gradient of the objective function is easy to determine. The only difficulty in this gradient calculation is to determine the gradient of the failure probability with respect to the design parameters $\mathbf{q}$, but this gradient is a by-product of the first-order reliability analysis.
- the gradient of the constraints involves the first but not the second derivatives of the limit state function.

The disadvantages of the nested optimization are:

- convergence problems may occur due to a difficulty in performing a proper scaling for the outer optimization.

- the computation time may be unnecessarily large as the reliability analysis calculation must be performed in each step of the outer optimization. Intuitively it does not seem necessary to carry the inner optimization to the end for the first steps in the outer optimization.
- no standard optimization program can be used for the complete optimization.

Instead of using the nested approach the two optimizations can be combined. The formulation becomes

$$\begin{aligned} &\textit{min. } C_T(\mathbf{p},\mathbf{q}) = C_I(\mathbf{p},\mathbf{q}) + C_F(\mathbf{p},\mathbf{q})\,\Phi(-|\mathbf{u}|) \\ &\textit{s.t. simple constraints on design parameters} \\ &\textit{s.t. } g(\mathbf{u},\mathbf{q})=0 \\ &\textit{s.t. } \frac{\mathbf{u}}{|\mathbf{u}|} = -\frac{\nabla_u g(\mathbf{u},\mathbf{q})}{|\nabla_u g(\mathbf{u},\mathbf{q})|} \end{aligned} \tag{7.7}$$

where $\nabla_u g()$ is the gradient vector of the g-function with respect to the u-variables. The last constraint can be rewritten as

$$\mathbf{u}^T\nabla_u g(\mathbf{u},\mathbf{q}) + |\mathbf{u}|\,|\nabla_u g(\mathbf{u},\mathbf{q})| = 0 \tag{7.8}$$

The advantages of this formulation are:

- a standard optimization program can be used
- scaling problems can be handled as done by standard optimization programs

The disadvantages of the formulation are:

- optimization routines which utilize the gradient of the constraints require the calculation of the second derivative of the limit state function
- the transformation from the physical variables to the standard normal u-variables must be included explicitly. This can be done by using modules from a reliability analysis program but requires some additional work.

For a complicated structural analysis problem the efficiency of the two approaches is measured by comparing the number of calls to the limit state function, since this calculation is the main time consuming operation.

7.1 SENSITIVITY ANALYSIS

It was mentioned earlier that the cost parameters are uncertain and it is of interest to determine the sensitivity of the solution of the optimization to changes in the cost function parameters.

First a simplified formulation is considered where the design parameters $\mathbf{q}$ are replaced by the reliability index β in the optimization problem. β is a function of $\mathbf{q}$. The total expected cost is

$$C_T(p,\beta) = C_I(p,\beta) + C_F(p)\Phi(-\beta) \tag{7.9}$$

where only a single cost function parameter p is considered. The minimum total expected cost is obtained for $\beta=\beta_t$, and

$$\frac{\partial C_T}{\partial \beta}\Big|_{\beta_t} = 0 \tag{7.10}$$

implicitly determines β_t as a function of p.

A change in p leads to a change in the minimum value for the total expected cost and in the solution for β. The change in the total expected cost is

$$\frac{dC_T}{dp} = \frac{\partial C_T}{\partial p} + \frac{\partial C_T}{\partial \beta}\frac{\partial \beta}{\partial p} = \frac{\partial C_T}{\partial p} \tag{7.11}$$

The change in the optimal reliability index is determined from

$$\frac{d}{dp}\left(\frac{\partial C_T}{\partial \beta}\right) = 0 \tag{7.12}$$

This leads to

$$\frac{\partial^2 C_T}{\partial p \partial \beta} + \frac{\partial^2 C_T}{\partial \beta^2}\frac{\partial \beta}{\partial p} = 0 \Rightarrow \frac{\partial \beta}{\partial p} = -\frac{\dfrac{\partial^2 C_T}{\partial \beta \partial p}}{\dfrac{\partial^2 C_T}{\partial \beta^2}} \tag{7.13}$$

calculated for $\beta = \beta_t$. These sensitivity factors are easy to compute.

Next the optimization in terms of the design parameters $\mathbf{q}$ is analysed. The total expected cost is

$$C_T(p,\mathbf{q}) = C_I(p,\mathbf{q}) + C_F(p)\Phi(-\beta(\mathbf{q})) \tag{7.14}$$

where again only one cost function parameter p is considered. The minimum total expected cost is determined for $\mathbf{q} = \mathbf{q}_t$.

$$\left.\frac{\partial C_T}{\partial q_i}\right|_{\mathbf{q}_t} = 0, \quad i = 1,2,\cdots,I \tag{7.15}$$

then implicitly determines $\mathbf{q}_t$ as a function of p.

A change in p leads to a change in the minimum value for the total expected cost and to the solution for $\mathbf{q}$. The change in the total expected cost is

$$\frac{dC_T}{dp} = \frac{\partial C_T}{\partial p} + \sum_{i=1}^{I}\frac{\partial C_T}{\partial q_i}\frac{\partial q_i}{\partial p} = \frac{\partial C_T}{\partial p} \tag{7.16}$$

The change in the optimal values for the design parameters is determined from

$$\frac{d}{dp}\left(\frac{\partial C_T}{\partial q_i}\right) = 0 \tag{7.17}$$

This leads to

$$\frac{\partial^2 C_T}{\partial p \partial q_i} + \sum_{j=1}^{I}\frac{\partial^2 C_T}{\partial q_i \partial q_j}\frac{\partial q_j}{\partial p} = 0; \quad j = 1,2,\cdots,I \tag{7.18}$$

This gives I linear equations for the derivatives $\partial q_i / \partial p$. The required second derivatives are

$$\frac{\partial^2 C_T}{\partial p \partial q_i} = \frac{\partial^2 C_I}{\partial p \partial q_i} + \frac{\partial^2 C_F}{\partial p \partial q_i}\Phi(-\beta) - \frac{\partial C_F}{\partial p}\phi(\beta)\frac{\partial \beta}{\partial q_i} \tag{7.19}$$

$$\frac{\partial^2 C_T}{\partial q_i \partial q_j} = \frac{\partial^2 C_I}{\partial q_i \partial q_j} + \frac{\partial^2 C_F}{\partial q_i \partial q_j}\Phi(-\beta) \tag{7.20}$$

$$- \frac{\partial C_F}{\partial q_i}\phi(\beta)\frac{\partial \beta}{\partial q_j} - \frac{\partial C_F}{\partial q_j}\phi(\beta)\frac{\partial \beta}{\partial q_i}$$

$$+ C_F \beta \phi(\beta)\frac{\partial \beta}{\partial q_i}\frac{\partial \beta}{\partial q_j} - C_F \phi(\beta)\frac{\partial^2 \beta}{\partial q_i \partial q_j}$$

The difficult term is $\partial^2\beta/(\partial q_i \partial q_j)$ as $\partial\beta/(\partial q_i)$ is given in (5.5). The second derivative can be expressed in terms of $\nabla_u g$, $\Delta_u g$, $\partial g/\partial q_i$, $\partial \nabla g_u/\partial q_i$, and $\partial^2 g/(\partial q_i \partial q_j)$. This derivative is thus somewhat complicated to compute.

8. Process Concepts In Structural Reliability

8.1 SOME MODELING CONSIDERATIONS

It was noted in section 2 at the definition of the physical formulation space, that time variation can be taken into account in an elementary way by discretizing to a finite number q of time points within a given period of time. Then the uncertain variation in time of some uncertain quantity is modeled by a vector $(X_1, \cdots, X_q)$ of q uncertain quantities corresponding to the q time points respectively. It is natural to generalize to a continuous time variation writing the uncertain physical quantity as $X(t)$ indicating that it is an uncertain function of time t. This is the concept of an uncertain process. Its interpretation is simply that for any finite set of time points $t_1, \cdots, t_q$ the corresponding set of values $(X(t_1), \cdots, X(t_q))$ is an uncertain vector. Thus we may have an expectation function (mean function) $E[X(t)]$ of t and a covariance function $Cov[X(t_1),X(t_2)]$ of (t_1,t_2).

It is outside the scope of this paper to present the fundamentals of uncertain processes (in the usual axiomatic set up for probability theory called "random stochastic processes" or "stochastic processes"). The reader is referred to the standard text books for an elementary presentation of the concepts of uncertain processes.

8.2 CROSSING THEORY IN STRUCTURAL RELIABILITY

The aforementioned methods of random process theory of particular relevance to structural reliability analysis are related to the study of the uncertain number of crossings per time unit of a trajectory out of a given suitable regular subset of space. This subset may be the safe set in the physical formulation space. If the trajectory for just one point in the time interval $[0,T]$ is outside the safe set, then there is failure at latest to time T and vice versa. Obviously this event occurs if the point of the trajectory corresponding to time zero is in the failure set or there is at least one outcrossing in $[0,T]$ of the trajectory out of the safe set. Let the probability of the first event be

$P_f(0)$ consistent with writing the probability of failure in $[0,T]$ as $P_f(T)$. The probability of the last event is at most equal to the expected value $E[N(T)]$ where $N(T)$ is the uncertain number of outcrossings in $[0,T]$. This statement is a simple consequence of the inequality $N(T) \geq I\,(N(T) \geq 1)$ where the right hand side is defined to be 1 if $N(T) \geq 1$ and zero otherwise. The expectation of the right side is the probability of the event $N(T) \geq 1$. Thus

$$P_f(T) \leq P_f(0)+E[N(T)] \tag{8.1}$$

For high reliability structures this upper bound is close enough to $P_f(T)$ to make $-\Phi^{-1}(P_f(0)+E[N(T)])$ a satisfactory lower bound approximation to the reliability index $\beta(T)=-\Phi^{-1}(P_f(T))$. Thus the reliability analysis reduces to the calculation of $P_f(0)$ (or $\beta(0)$) by use of FORM/SORM reliability analysis described herein, and calculation of the expected number of outcrossings in $[0,T]$.

Before leaving the subject for now, it should be noted that the right hand side of (8.1) in practice often is simplified either by neglecting the second term, which amounts to neglecting the effect of time variation, or neglecting the first term. This is simply because the reliability index is very insensitive to variations of the failure probability given that the reliability index is reasonably large. For example, if the neglected term is less than half of the other term, then the error of the reliability index is less than about 5% if the reliability index is larger than 3.5.

The expected number of outcrossings is often computed by Rice's formula or a generalization of this to a multidimensional process. A conceptually simple alternative formulation uses the sensitivity factor for a parallel system. Let the now time varying safety margin $M(t)$ be a function of the stochastic vector process $\mathbf{Z}(t)$. The time derivative of the stochastic safety margin is denoted as $\dot{M}(t)$. The mean rate of crossings into the failure set is

$$\nu = \lim_{\Delta t \to 0} \frac{1}{\Delta t} P(M(t)>0 \cap M(t+\Delta t<0)) \tag{8.2}$$

When the safety margin is differentiable this becomes

$$\nu = \lim_{\Delta t \to 0} \frac{1}{\Delta t} P(M(t)>0 \cap M(t)+\dot{M}(t)\Delta t<0) = \frac{d}{dp} P(M>0 \cap M+\dot{M}p<0)_{p=0} \tag{8.3}$$

where the last derivation is straightforward and shows that the outcrossing rate can be computed as the sensitivity factor for a parallel system.

9. Discussion

Some fundamental concepts of reliability analysis and reliability assessment have been described. Modeling of inherent uncertainty, statistical uncertainty and model uncertainty is described and all sources of uncertainty are included in the analysis causing a traditional relative frequency interpretation of the failure probability invalid. The most promising methods for reliability calculations are the first- and second-order reliability methods and various efficient simulation methods. These methods provide important sensitivity and importance factors in addition to a failure probability. Finally the methods are shown to be efficient also for reliability updating, reliability based design and reliability evaluations involving random processes.

The reliability theory is now well established and accepted, and efficient numerical computer analysis tools are available making the methods accessible for a large group of practicing engineers. A model code has been proposed for direct use of reliability methods in structural design and a more international code within this direction is anticipated increasing the use of reliability methods in every day work.

10. General Bibliography - books on reliability methods and codes and model codes

1. Ang, A.H.-S. and Tang, W.H., *Probability Concepts in Engineering Planning and Design,* Vol. I&II, John Wiley, 1984.
2. Augusti, G., Baratta, A. and Casciati, F., *Probabilistic Methods in Structural Engineering,* Chapman and Hall, 1984.
3. Benjamin, J.R. and Cornell, C.A., *Reliability, Statistics and Decision for Civil Engineers,* Mc-Graw Hill, 1970.
4. Bolotin, V.V., *Wahrscheinlichkeitsmetoden zur Berechnung von Konstruktionen,* VEB Verlag fur Bauwesen, 1981.
5. Borges, J.F. and Castanheta, M., *Structural Safety,* Laboratorio Nacional de Engenharia Civil Lisbon, 1971.
6. Ditlevsen, O., *Uncertainty Modeling with Applications to Multidimensional Civil Engineering Systems,* Mc-Graw Hill, 1981.
7. Ditlevsen, O. and Madsen H.O., *Structural Reliability Methods (in danish),* SBI, 1990.
8. Madsen, H.O., Krenk, S. and Lind, N.C., *Methods of Structural Safety,* Prentice-Hall, 1986.
9. Melchers, R.E., *Structural Reliability and Predictions,* Ellis Horwood/J. Wiley, 1987.
10. Schueller, G., *Einfuhrung in der Sicherheit und Zuverlassigkeit von Tragwerken,* Verlag Wilhelm Ernst & Son, Berlin, 1981.
11. Thoft-Christensen, P. and Baker, M., *Structural Reliability Theory and Its Applications,* Springer Verlag, 1982.
12. Thoft-Christensen, P. and Murotsu, Y., *Application of Structural Systems Reliability Theory,* Springer Verlag, 1986.
13. American Institute of Steel Construction (AISC), "Load and Resistance Factor Design Specifications," AISC, Chicago, 1986.
14. American National Standards Institute (ANSI), "American National Standard Minimum Design Loads for Buildings and other Structures," ANSI A58.1, ANSI, New York, 1982.
15. CEB, "First Order Concepts for Design Codes," CEB Bulletin No.112, Munich, 1976.
16. CEB, "Common Unified Rules for Different Types of Construction and Material, Vol. I," CEB Bulletin No.116, Paris, 1976.
17. CIRIA, "Rationalisation of Safety and Serviceability Factors in Structural Codes," CIRIA Report No. 63, London, 1977.
18. ISO, "General Principles on Reliability for Structures," ISO/DIS 2394, 1984.
19. NBS, "Development of a Probability Based Load Criterion for American National Standard A58," U.S. Department of Commerce, NBS Special Publication 577, 1980.
20. Nordic Committee for Building Structures (NKB), 'Recommendation for Loading and Safety Regulations for Structural Design,' NKB Report No. 36, 1978.
21. JCSS, "Proposal for a Code for the Direct Use of Reliability Methods in Structural Design", O. Ditlevsen and Henrik O. Madsen, Working Document, 1989.

RELIABILITY ASSESSMENT OF MULTI–MEMBER STRUCTURES

R. Rackwitz
Technical University of Munich
Arcisstr. 21, 8000 Munich 2, F.R.G.

1. Introduction – Historical Remarks

Traditionally structural safety considerations are based on checking safety at the component level. Only implicitly, adjustments of the required safety margins with respect to the system reliability aspect have been made in some cases. For example, in stability problems in soil mechanics the representative values of the soil properties are taken close to their means in order to take account of the averaging of the property fluctuations along the yield surfaces the sections of which are to be viewed as a parallel system. However, that almost all structures involve a system reliability aspect must be considered as common knowledge. It is recognized that some systems behave like series systems where system collapse is implied by the failure of a component in any of its failure modes. Such systems are correctly categorized as especially critical systems. It is also recognized that many civil engineering structures are redundant structures in the sense that given the (first) failure(s) of some component(s), element(s), joint(s), cross section(s), etc., the structural system not necessarily fails, too. Without such inherent redundancy many of damaged structures would actually have failed. Thus, redundancy has been identified as a desirable feature of a structure although it usually does not mean that the user will leave a structural component unrepaired once it failed. Vague notions like robustness have been introduced as requirements for structural design meaning essentially that after the failure of one component in an extreme loading event the structure still maintains its primary function though with somewhat reduced capability. And it is not surprising that already in the very early deterministic works on the subject the ductility of structural components has been identified as one of the dominating factors when assessing system reliability.

A systematic study of the probabilistic problem of structural system reliability has not started earlier than about 1965 when formulations where presented which carried the potential of becoming a theory (see, for example, Shinozuka/Itagaki, 1966; Moses/Kinser, 1967; Jørgensen/Goldberg, 1969; Yao/Yeh, 1968; Stevenson/Moses, 1970; Vanmarcke, 1973). Most of those studies were based mechanically on the simple rigid–plastic model or on perfect elastic brittleness. It turned out however that the computational tools available where not sufficient to deal with real structural systems both from a mechanical and a probabilistic point of view. For the probabilistic part it

J. Bodig (ed.), Reliability-Based Design of Engineered Wood Structures, 47–73.

was found that the so–called second–moment methods where not rich enough and numerical or Monte Carlo integration required far too much numerical effort. Notwithstanding the fact that important general insights have been gained during the seventies the reliability of systems became a serious research subject only since about 1980. The directions of development were twofold. On the one hand the computation methods for probability integrations could substantially be improved and, on the other hand, important precision could be introduced in the formulations at the mechanical side. It is fair to say that considerations concentrated on a few somewhat academic structures such as the chain with n links, the rigid–plastic portal frame or its slightly more complicated derivates and the bundle–of–threads system with equal load–sharing. Those systems also appear to be the ones from which it is most easy to draw some more general conclusions. They belong to the set of structures where extra work would contribute rather little to what is already known.

In this paper the present status of the reliability of quasi–static structural systems is reviewed. This review will focus on the results of engineering relevance rather than on methods which have already been reviewed thoroughly by Ditlevsen/Bjerager (1986). The computational tools available will be discussed. It will be explained why time–variant system reliability problems are not yet covered by this paper. It will be found that for series systems a well developed theory is available. Then, the so–called Daniels system, an ideal parallel system, will be studied on the basis of a previous paper on this subject by the author (Gollwitzer/Rackwitz, 1990). It serves to gain some general insights. An overview of the different approaches to structural systems of general topology and with general mechanical properties will also be given with special emphasis to the so–called failure tree approach which at present is the most common and general. The author apologizes for a slight bias in this review to developments where he was personally involved. Also, the review only covers publications up to early 1990.

2. Computational tools

It is necessary to first discuss the state of computational tools. Today some powerful formal tools for the computation of systems as represented by the union of cut sets of componental failure events are available where it is important to note that componental behavior is modeled by a simple Boolean representation (see, for example, Hohenbichler/Rackwitz, 1983 for the basic results in first–order reliability, Hohenbichler et al., 1987, and Breitung/Hohenbichler, 1989 for the basic second–order results and Bjerager, 1990, for a review of alternatives based on important sampling techniques). It will be demonstrated to some extent that even with these tools in hand there is still a long way to go to compute the reliability of realistic, redundant structural systems.

While for quasi–static, non–deteriorating systems under time invariant loading the capabilities of modern FORM/SORM methods or other approaches are sufficient, problems will already be met when the loads vary in time and additional complication will be occur if the system properties deteriorate in time either by load independent actions or by load induced fatigue. Then random process theory has to be applied. So far, the stationary case is most developed. For Gaussian vector process loading an important result is by Ditlevsen (1983) for the crossing rate out of the (linear or linearized) failure domain of a series system. This result can be applied with some less generality to rigid plastic redundant systems. Some further considerations have been given for the same type of loading to the crossings out of unions of intersection by Giannini/Nuti/Pinto (1985) and Schrupp/Rackwitz (1985). The crossing rates for Gaussian vector processes into intersections of failure domains were determined by Hohenbichler/Rackwitz (1986) in a second order context. Wen/Chen (1986) considered certain systems under combinations of extremely filtered rectangular wave pulse loadings. The author determined the crossings of rectangular wave renewal processes into unions of intersections of (linear or linearized) failure domains (Rackwitz, 1985). Although no result for other processes used to model structural loads is known to the author it is probably not too difficult to proceed along the lines described in the mentioned papers for some other load processes of interest.

For the non–stationary case series systems appear to be manageable by straightforward generalizations of the stationary case but no result is known at present for parallel systems and one can only presume that results are most easily obtained for appropriately linearized failure domains. For redundant structures behaving dynamically and/or where the degradation of redundancy by subsequent or joint failure of the components produces dynamic effects only a very few exploratory results are available (see, for example, Guers/Dolinski/Rackwitz, 1988 and Guers/Rackwitz, 1987). Any generalization of these results to realistic redundant systems will require enormous computation effort. Some further results have been obtained by Grigoriu (1990) for the bundle–of–threads system.

Thus it is concluded that practical system reliability studies still must be restricted to cases where the fairly well developed methods for probability integrations are applicable. This implies that the modeling of loads and system properties must be possible by simple random vectors.

3. Series systems

In a review of the methods and the results available for time–invariant structural systems simple series systems must be considered first. Not only are the computational tools ready for practical applications. For series systems it is also not necessary to distinguish between different types of mechanical behavior of the components. But most important, any other multi–component, multi–mode system can be treated as a series system whose analysis provides a more or less satisfying upper bound to the system failure probability. It is noteworthy remarking that according to Henley/Kumamoto (1981) the series system problem with the German missiles V2 in the forties is said to be the first reliability problem which was attacked by probabilistic methods. For series systems the failure probability can easily be bounded from below by the largest componental failure probability and from above by the sum of all componental failure probabilities. When these so–called first order bounds are satisfactory all time–variant problems which have a solution on the component level also have a solution on the system level. In the context of asymptotic SORM it has been shown that the upper bound is even asymptotically exact (for example, Breitung, 1984, for the time–invariant case and Leadbetter et al., 1983, for time–variant case). The sum over the componental failure probabilities then needs to be extended only over those components which have smallest (and equal) safety indices. Exact probabilities are difficult to obtain but the mentioned bounds can be efficiently improved by using a technique proposed by Ditlevsen (1979) and others. These so called second–order bounds involve the probabilities of joint failures of any two components which can be computed by

FORM/SORM. Only for low reliability systems it can be worthwhile to improve those bounds by third order terms (see for details Hohenbichler/Rackwitz, 1983 and Ramachandran/Baker, 1985). Another way to compute series systems is to pass over to the corresponding parallel system for the safe sets of structural states. Those sets usually are convex sets. If each set boundary is approximated either by a linear or quadratic form it is again possible to use FORM/SORM. The FORM result then corresponds to the computation of the multinormal integral for which efficient methods are available (see, for example, Hohenbichler/Rackwitz, 1986; Gollwitzer/Rackwitz, 1988). Finally, it is worthwhile to mention that series systems have, by definition, an extremely useful property which is independence of its reliability from the load path. If this independence is not present as we will see when discussing redundant systems serious problems in modeling and computation can occur.

4. Daniels systems

The other extreme system is the ideal bundle–of–threads system, a system which has the largest possible degree of redundancy. Because it has attracted the attention of many researchers from various fields and with different background and interests during the last 47 years its reliability properties are well known under quite different circumstances. Therefore, it will be discussed in some detail with particular reference to more general insights in the system reliability aspects of redundant structures.

Daniels (1945) assumed independently, identically distributed componental strengths X_i ($i=1,...,n$), a constant modulus of elasticity and perfect so–called equal load–sharing among the unbroken fibers (components, elements, members,...). Furthermore, the fibers behave ideally brittle. The load acts quasi–static. Obviously, the system strength R_n under these conditions is

$$R_n = \max_{i=1}^{n} \{(n-i+1)\,\hat{X}_i\} \tag{1}$$

where the $\hat{X}_i$'s are the ordered elemental strength values X_i such that $\hat{X}_1 \leq \hat{X}_2 \leq ... \leq \hat{X}_n$. The system failure probability under load $S = s$ becomes:

$$\begin{aligned} P_f &= P(R_n \leq s) = P\left(\bigcap_{i=1}^{n} \{(n-i+1)\,\hat{X}_i - s \leq 0\}\right) \\ &\leq \min_{i=1}^{n} P(\{(n-i+1)\,\hat{X}_i - s \leq 0\}) \end{aligned} \tag{2}$$

The second line corresponds to the "strongest" component and represents an upper bound for parallel systems which, unfortunately, is rather conservative in most cases.

Daniels not only found a recursive scheme for the determination of the probability distribution of system strength which later could be rearranged by several authors to improve its numerical performance. It still is the only compact, exact solution known for a redundant structural system with non–perfectly ductile elements. For convenience, a numerically suitable scheme is given here

$$P(R_n \leq s) = \{1 - \bar{F}(s)\}^n - \sum_{k=0}^{n} \begin{bmatrix} n \\ k \end{bmatrix} S_k \left\{\bar{F}(\tfrac{s}{n-k}) - \bar{F}(s)\right\}^{n-k}$$

with

$$S_0 = 1$$

$$S_m = \left\{1 - \bar{F}(\tfrac{s}{n-m-1})\right\}^m - \sum_{r=0}^{n} \begin{bmatrix} n \\ r \end{bmatrix} S_r \left\{\bar{F}(\tfrac{s}{n-r}) - \bar{F}(\tfrac{s}{n-m-1})\right\}^{m-r}$$

and $\bar{F}(x)$ the complementary distribution function of strenght X. Daniels also found a simple formula for system strenght when n becomes very large which will be discussed below in some detail.

The contrary case of ideal plasticity is trivial because the system strength R_n then is just the sum of componental strengths. For large systems, the central limit theorem holds for the distribution of system strength under suitable conditions (Liapunov conditions).

Unfortunately the system is a realistic model only in a few cases as, for example, for parallel wire cables, for fiber–reinforced composite materials with a soft matrix and for certain fastening structures.

4.1 LARGE DANIELS SYSTEMS

Daniels derived the Gaussian distribution as a limiting distribution of system strength for $n \to \infty$. Because his result is considered as so important it is given here for easy reference. The component strength is distributed continuously as $F_X(x)$ with $x > 0$ and $F_X(0) = 0$. Then, Daniels' asymptotic result is:

$$\lim_{n \to \infty} P(R_n \leq s) = \Phi\left(\frac{s-E_n}{D_n}\right) \tag{3}$$

with

$$E_n = n\, x_0\, (1 - F_X(x_0)) + C_E \tag{4}$$

$$D_n^2 = x_0^2\, [n F_X(x_0)(1 - F_X(x_0))] + C_D \tag{5}$$

and x_0 the maximum point of $x\,(1 - F_X(x))$ or the unique solution of

$$\frac{\partial}{\partial x} [x(1 - F_X(x))] = 0 \tag{6}$$

C_E and C_D are correction terms to be discussed below. If, in particular, the X_i are Weibull–distributed according to $F_X(x) = 1 - \exp[-\lambda x^\beta]$ one determines $x_0 = (\lambda\beta)^{-1/\beta}$. For most other distribution functions eq. (6) has to be solved numerically. The basic result is best interpreted by observing that $(1 - F_X(x))$ is the proportion of unbroken fibers at level x and $nx\,(1 - F_X(x))$ is their minimum strength in the sense that larger strengths than x are set equal to x. Eq. (6) maximizes this strength. But the actual number of fibers for given proportion $F_X(x)$ of failed members is random. It is binomially distributed and according to the central limit theorem in the Moivre–Laplace version asymptotically Gaussian. We first note that system strength is always smaller than average strength. For the Weibull distribution of X system strength depends primarily on the shape (dispersion) parameter β of this distribution. Approximately, there is $\beta \approx 1.2/V$ with V the coefficient of variation of X. Therefore, the critical strength x_0 increases with decreasing coefficient of variation and consequently the proportion of unbroken fibers also increases with decreasing coefficient of variation. This type of dependence on the underlying distribution function of X can also be observed for other distribution functions.

Later several statisticians became interested especially in the asymptotics of such systems and in weakening the rather restrictive assumptions originally made by Daniels on the stochastic model. In particular, convergence towards the asymptotic result was observed to be rather slow. Daniels himself obtained an improvement for the mean of the limiting distribution maintaining the original assumptions on the stochastic and mechanical model (Daniels, 1974). The correction term C_E by Daniels using quite different statistical arguments is:

$$C_E = 0.966\, n^{1/3} a \tag{7a}$$

where

$$a^3 = f_X^2(x_0)\, x_0^4 / (2\, f_X(x_0) + x_0\, f_X'(x_0))$$

Barbour (1981) improved the variance of the limiting distribution but his improvement is numerically less important

$$C_D = 0.317\, n^{2/3}\, a^2 \tag{7b}$$

Nevertheless, those improvements yield good results already for moderate n.

Sen/ Bhattacharyya (1976) found that a limiting Gaussian distribution is also obtained if the strengths of the components are dependent but fulfill certain mixing conditions. Later their result was generalized to so–called continuous systems, i.e. where the strength is a continuous, homogeneous process or field over a certain domain (Hohenbichler 1983). Unfortunately the variance of system strength then is not so easily computed. Hohenbichler/Rackwitz (1981) introduced a special type of strong dependence and found that Gaussianity will not be reached in this case. For example, let the fiber strength be given by $X_i := X_0 X_i$ where X_0 is a variable with given distribution which is common to all fibers. Then, the distribution of R_n will be dominated asymptotically by the distribution of X_0.

An important mechanical generalization was achieved by Phoenix/Taylor (1973). Their basic idea was to reformulate the problem in terms of deformations rather than strength. They kept the independence assumption but adopted relatively general force–deformation curves for the properties of the components and again were able to prove a limiting Gaussian distribution for the system strength. In particular, they assumed that not only the breaking strengths of the fibers are random but also their deformation at rupture. In yet more generality, the force–deformation curve can be non–linear. Phoenix/Taylor still assumed that the fibers break brittle even after some plastic deformation. But a smooth force–deformation curve can also be assumed (Hohenbichler/Rackwitz, 1983). It is a simple exercise to show that a critical deformation exists where system strength becomes maximal. The corresponding strength is the system strength. It is further possible to determine its variance numerically. Although it turns out to be difficult to show that system strength is normally distributed there is no doubt that it is so when the componental properties are sufficiently mixing. In the same paper by Phoenix/Taylor (1973) the theory is also generalized to fibres experiencing random slip or slack at the clamps. An interesting discussion and some further extensions of the general case has been given recently by Daniels (1989).

There has been some discussion whether the Gaussian distribution also holds in the extremes of system strength. Some theoretical and numerical results indicate that the tails of the distribution of system strength are probably better approximated by an extreme value distribution but strong theoretical justification so far is missing.

Further results have been obtained by relaxing the assumption of equal load–sharing in the system. Harlow/Phoenix (1981) developed not only results for so–called local load–sharing rules based on the earlier proposals of Rosen (1964) and Zweben (1968) but also extended the theory to chains–of–bundle systems (see also Gücer, 1962, Smith/Phoenix, 1981). In both cases, the limiting distribution now becomes an extreme value distribution. Local load–sharing rules (stress concentrations around broken fibers) imply a tendency to progressive failure once the weakest fiber(s) is (are) broken and, therefore, the Weibull–distribution is a natural candidate of system strength given that componental strength has a power law behavior of $F_X(x)$ in the lower tail. Then, the asymptotic strength of a chain–of–bundle system should also be Weibull–distributed. This could, in fact, be shown and we refer to Smith (1983) for a review and a discussion of the relevant literature. If, on the other hand, there is equal load–sharing in the bundles the asymptotic strength is Gumbel distributed (Smith/Phoenix, 1981). The latter result appears less convincing and it certainly will have to be modified if stronger results become available for the distribution of system strength in its lower tail which also take account of the fact that system strength can only be positive.

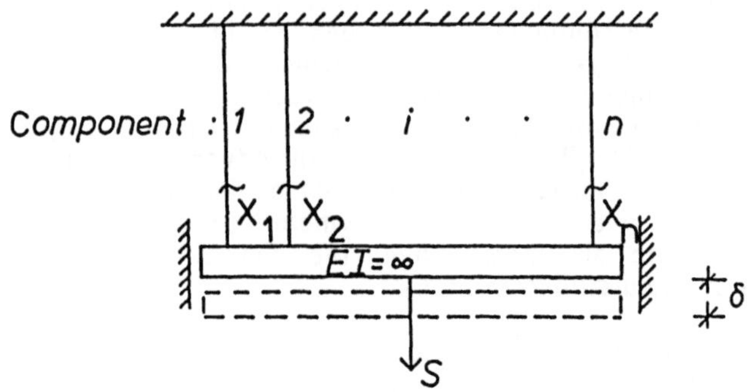

Fig. 1. Daniels system

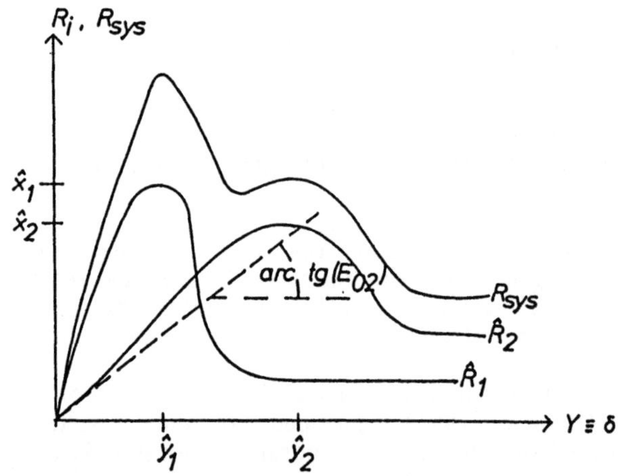

Fig.2. Typical componental force–deformation curves

The important aspect is that whenever there is a load–sharing with sufficient concentration the system tends to behave as a series system rather than a parallel system. Analogous trends are to be expected when the load–sharing regime is deterministically non–uniform.

Whereas asymptotic concepts appear to be appropriate when deriving statistical strength models because n can be assumed to be large, this is rather questionable for small systems. As mentioned, convergence to the asymptotic distribution is rather slow even in Daniels' original classical case (see Smith/Phoenix, 1981 for some interesting theoretical observations concerning the rate of convergence and the tails of the asymptotic distribution and Hohenbichler/Rackwitz, 1983 for numerical comparisons).

4.2 NUMERICAL RESULTS FOR SMALL DANIELS SYSTEMS

Quite another route of research focused on small to medium systems with the intention to study structural redundancy for non–ductile materials in general. Daniels' original model actually appears to be very attractive for this purpose due to its simple mechanics. From a probabilistic point of view Daniels' model, however, is rather complicated because of its many possible, equally important sequences of componental failures to system collapse. For this reason the Daniels system has frequently been used as a demonstration example in structural system reliability studies and we will mention a few references of direct interest in the context of this paper. For example, Shinozuka/ Itagaki (1966) found the transition probability from one state of the system into the next. Kersken–Bradley (1981) applied the model to structural timber concentrating primarily on a description of system strength by its first and second statistical moments and the development of strength under imposed deformations. Hohenbichler/Rackwitz (1983) presented a formulation amenable to the application of modern first– and second–order reliability methods. It is based on the so–called order–statistics approach used in eq. (1) and which has repeatedly been used later on (see also Rackwitz/Hohenbichler, 1981 and Hohenbichler et al., 1981, where certain special but practically important cases are treated such as random shapes of the componental force–deformation curves, "slip" or "slack" in the anchorage of the components, the existence of defect components, etc). Gollwitzer (1986) carried out numerous numerical studies with particular reference to the circumstances in redundant fastening systems. The remainder of this section concentrates on formulations and results for not too large Daniels systems with an equal load–sharing regime and whose components do not deteriorate with time. Furthermore, dynamics can be neglected except, possibly, for the phase of load redistribution after componental failure. Numerical results for this system with fairly general mechanical and stochastic characteristics for the components will be used to draw some general conclusions on the effect of redundancy on structural reliability. A numerically feasible formulation is particularly simple if the stochastic dependence between the different components can be described by a common set of variables. This type of dependence may be called "equi–dependence". Then, the order–statistics approach as in Hohenbichler/Rackwitz (1983) is still applicable after some modification (see Guers/Rackwitz, 1987) in conjunction with modern FORM/SORM techniques for the determination of the resulting probability integrals.

For an arbitrary force–deformation curve as in figure 2 the componental failure event for a given imposed deformation δ is:

$$F(\delta) = \{\sum_{i=1}^{n} R_i(\delta) - S \leq 0\} \tag{8}$$

Herein S denotes the possibly uncertain load and $R_i(\delta)$ denotes the uncertain componental force at deformation δ. System failure occurs if the maximum system resistance is exceeded by the load which can be described as follows:

$$F_{sys} = \{\max_{(\delta)} (\sum_{i=1}^{n} R_i(\delta)) - S \leq 0\} = \{\bigcap_{(\delta)} (\sum_{i=1}^{n} R_i(\delta) - S \leq 0)\} \tag{9}$$

Figure 2 gives typical force–deformation curves of components together with the curve of system resistance $R_{sys} = \Sigma R_i$. For a finite number of components it is always possible to define the order statistics $(\hat{Y}_1,...\hat{Y}_n)$ for the vector $(Y_1,...Y_n)$ of deformations where the components reach their maximum bearing capacity X_i. Formula (17) can be rewritten as:

$$F_{sys} \leq \{\max_{i=1}^{n} (R_i(\hat{Y}_i,Q_i) + \sum_{k\neq i}^{n} R_k(\hat{Y}_i,Q_k) - S \leq 0\} \tag{10}$$

where the Q_i's denote random vectors describing further properties of the force–deformation curve. The inequality sign now reflects the possibility of a larger system resistance for deformation states in between $[\hat{Y}_i, \hat{Y}_{i+1}]$ for i = 1,..., n–1. For the following parameter studies the components of the system are designed for a reliability index of $\beta_k = 2.0$ as if no system effect exists.

Ideal brittleness and ideal plasticity in independent components as limiting cases

The foregoing formulation can easily be specialized to the case of ideal brittleness and ideal plasticity. Clearly, ideal brittleness together with a linear elastic behavior in non–failed componental states corresponds to the least extra reliability provided by redundancy. On the contrary, ideal plasticity provides the largest extra reliability achievable by redundancy. These statements hold for all possible dependencies between the variables characterizing the components.

Figure 3 first demonstrates the influence of the mechanical behavior of the components on system reliability for independent properties of the components. In this figure showing the system reliability index versus the number of components two limiting curves are also included, corresponding to an ideal series system and an ideal parallel system and whose failure events are given below for easy reference.

$$F_{sys\cdot series} = \bigcup_{i=1}^{n} F_i \qquad F_{sys\cdot parallel} = \bigcap_{i=1}^{n} F_i$$

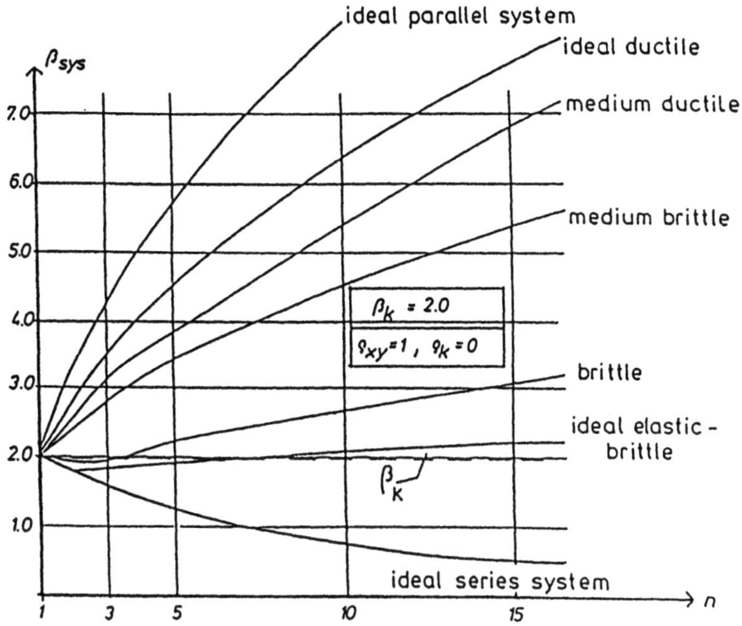

Fig. 3. System reliability index versus number of components for different force–deformation curves

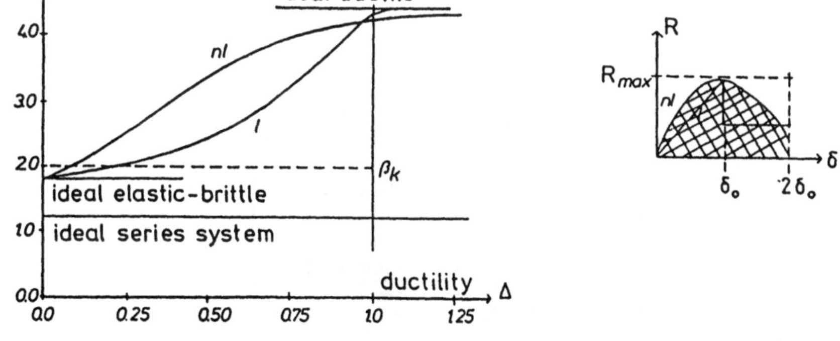

Fig. 4. System reliability index versus ductility

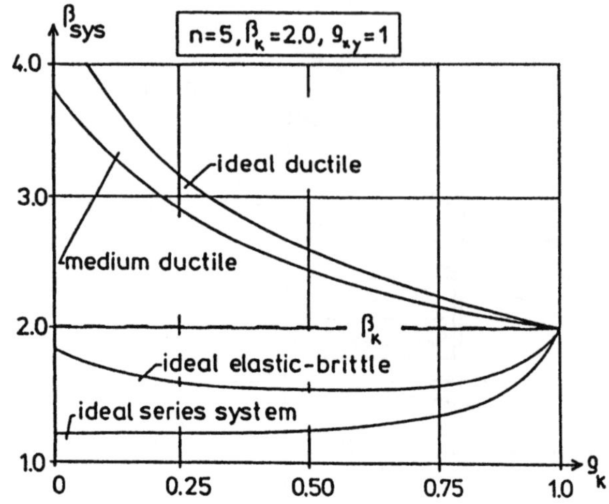

Fig. 5. System reliability index versus correlation ρ_k of strength between components

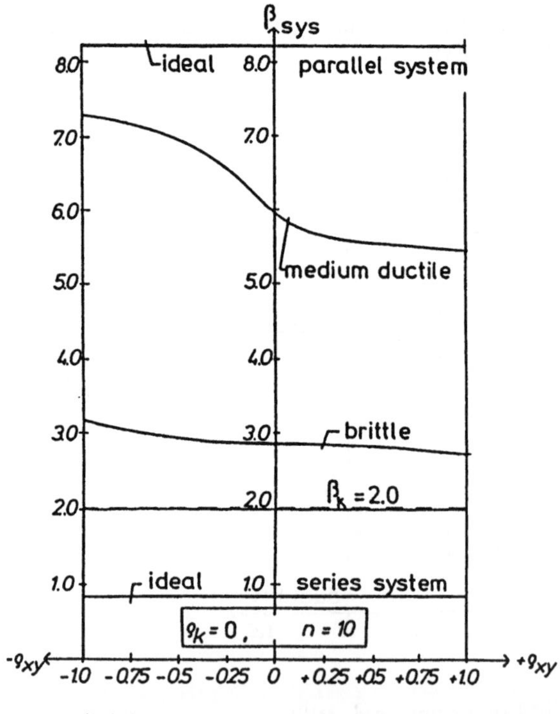

Fig. 6. System reliability index versus correlation ρ_{xy} between maximum strength and corresponding deformation between components

Note that the ideal parallel system with so–called hot redundancy has no plausible mechanical interpretation. The three other curves correspond to different degrees of ductility (see below). As expected, the reliability of the series system decreases with n. The reliability of the parallel system increases significantly with n. For the brittle system one observes first a decrease of reliability below the reliability level of a single component. Only for a larger number of components this level is exceeded by the system reliability and a significantly larger degree of redundancy is necessary to produce higher reliabilities but at a much smaller rate than for the more ductile systems. Obviously, for a small number of components the brittle Daniels system behaves like a series system.

This is interpreted as follows: if the weakest component breaks, it is unlikely in a small system that there are resistances of the remaining components which can carry the extra redistributed load. This means that a small degree of redundancy does not necessarily produce extra reliability in systems with brittle components. It is better to design a single but strong component to carry the load rather than to distribute its material among several components. For larger componental variabilities this unfavorable "series system" effect vanishes, however.

Influence of the degree of componental ductility

Next the effect of the degree of ductility is studied. We introduce the following measure of ductility

$$\Delta = \left(\int R(\delta)\, d\delta - \frac{R_{max} - \delta_0}{2}\right) / (R_{max}\, \delta_0) \tag{11}$$

It is further assumed that the maxima in the componental force deformation curves are fully correlated with the deformation at that point. Figure 4 then shows for a five–component–system the increase of reliability with ductility. It is recognized that the increase in the reliability index is roughly linear with ductility implying an exponential decrease of the corresponding failure probability up to relatively high ductilities around $\Delta = 1$ beyond which the conditions for the fully plastic case prevail. One can conclude that already relatively little ductility will provide considerable extra reliability. It is easily visualized and can be demonstrated numerically that for smaller ductilities this positive effect, however, is true only if the variability of the deformation at the maximum force is small (see above).

Stochastic Dependencies

The influence of stochastic dependencies within the variables characterizing the force deformation curve of a component and between components is investigated next. For example, if full correlation is still maintained between maximum force and deformation but a non–zero correlation coefficient is assumed for the maximum forces in the components, one can easily produce figure 5. It shows that the redundancy effect is largest for zero correlation and ideal ductility and vanishes for $\rho = 1.0$. For the elastic brittle case one observes that medium positive correlations can make the situation even worse for small systems. Again any effect of redundancy vanishes for full correlation. The reliability of the series system increases with correlation. In figure 6 one can see that correlation of the properties within a component appear to have relatively little importance at least for brittle systems.

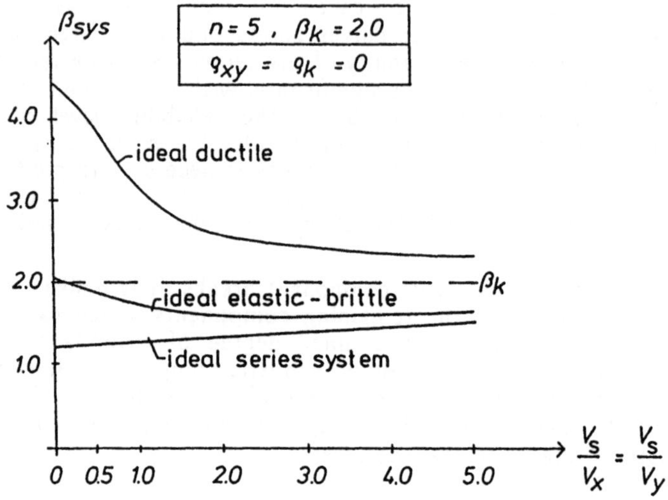

Fig. 7. System reliability index versus ratio of load variability to strength variability

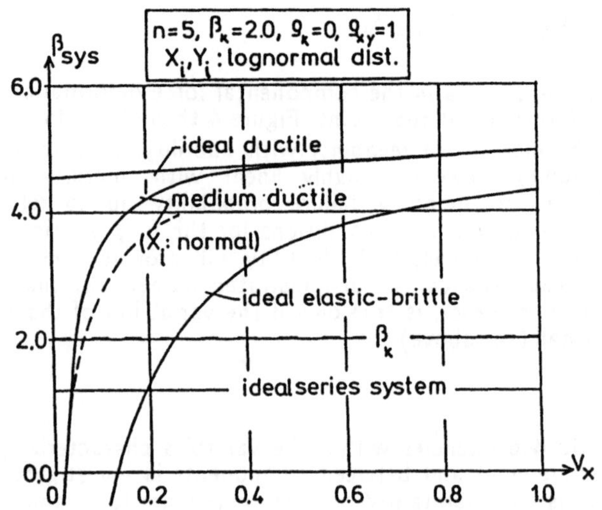

Fig. 8. System reliability index versus componental strength variability

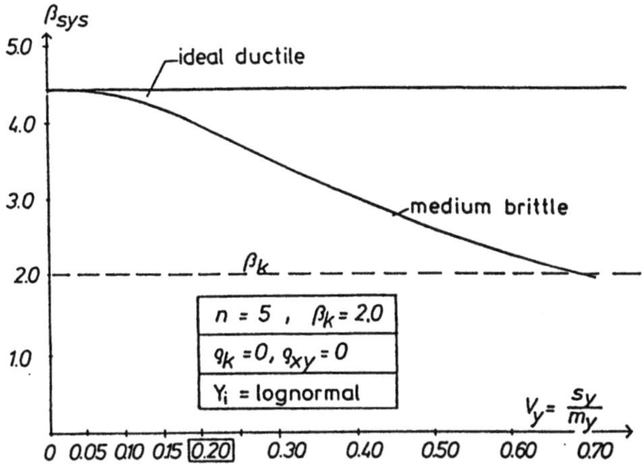

Fig. 9. System reliability index versus coefficient of variation of deformation at maximum componental strength

Dynamics of load–effect redistribution

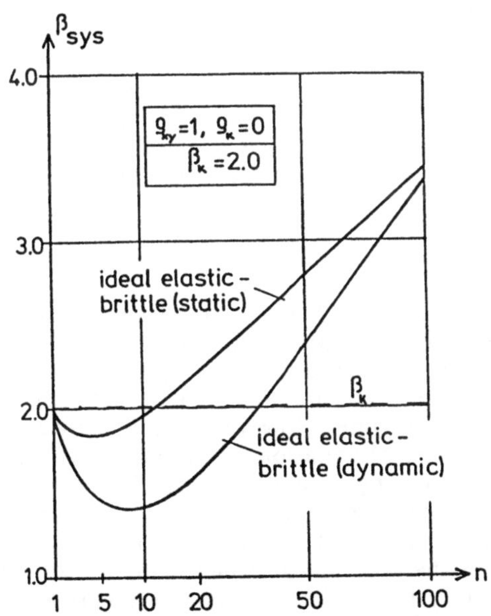

Fig. 10. System reliability index versus numbers of components with and without dynamic effects during redistribution

Relative Influence of Load and Resistance Variability

It is interesting to investigate the effect of different ratios of the coefficients of variation for the loads and the componental resistance. In view of figure 5 one expects a substantial decrease of reliability with increasing dominance of the load variability. This is, in fact, true as can be seen from figure 7, where the extra reliability due to redundancy reduces to an insignificant amount for ratios of the coefficients of variation V_S/V_X larger than two, say. The same type of investigation also allows to quantify the relative effect of the coefficient of variation of the componental resistances. In figure 8 the coefficient of variation V_X is varied ($V_S = 0$). One observes that an increasing coefficient of variation has a positive effect on the extra reliability for both the ductile and the brittle system. For the brittle system one can see that the unfavorable effect of redundancy for small systems is less pronounced or non–existing. Under these circumstances, the remaining components in a Daniels system are likely to have sufficient extra strength to avoid progressive collapse once the weakest component fails. Of course this does not mean that a larger strength variability is generally better than small variability because the larger variability has to be compensated for by a larger mean in structural design.

Another important factor for brittle systems is the coefficient of variation of the deformation at maximum strength. From figure 9 one concludes a dramatic decrease in reliability with V_Y. It simply means that for large V_Y it is unlikely that the components develop their (maximum) strength at about the same level of deformation. Also, the system force–deformation curve becomes rather flat.

Finally, the dynamic effect during load redistribution is quantified in figure 10 for a system with elastic brittle components following the limiting considerations in Guers/Rackwitz (1987). It is seen to be significant even for relatively large systems. Fujita et al. (1988) could show that the adverse dynamic effects only diminish significantly for relatively large damping of about 10% in the system.

Summary of results for the Daniels system

The above results and some further parametric studies not presented herein allow to draw some general conclusions. Hereby, it is important to bear in mind that among all different types of redundant structural systems the Daniel system as defined before must be considered as the system where structural redundancy produces most extra reliability due to its equal load–sharing regime. However, if the correlation between components is high and/or the coefficient of variation of the load exceeds the one of strength significantly, the gain in reliability by redundancy soon becomes insignificant. If, on the other hand, there is insignificant correlation between components and the coefficient of variation of loads and resistances are comparable in magnitude, the amount of extra reliability depends on the mechanical behavior of the components. For elastic brittle behavior there is relatively little effect for small systems and the largest relative gain in reliability can be achieved for large coefficients of variation of strength.

For small brittle systems there is even a negative effect of redundancy for small coefficients of variation. The Daniels system then behaves almost like a series system. In small to medium size brittle systems the dynamic effects during load–effect redistribution must be expected to be non–negligible and can reduce reliability quite significantly unless relatively large damping in the system reduces the dynamic effects to a sufficient degree. Only if the components behave ductile and their properties are weakly dependent there can be a considerable increase in reliability with the number of components.

5. General structural system reliability

Topologically arbitrary structures are much more difficult to discuss. Yet, a number of concepts and methods with their variants suitable for special cases exist. Due to the limited capability of present reliability methods a number of idealizations have to be made which primarily concern the mechanical model of the system, the failure criteria and the load model. If one restricts consideration to structures which can be handled in the frame work of FORM and SORM, i.e. where all uncertainties are represented by a random, possibly high–dimensional vector, one of the crucial idealizations is the discretization of the structure into a finite set of components or members. Those structural members generally are also the "finite elements" of the structure. In such a system the local resistance quantities and the local geometry can be treated as uncertain but the geometry and the load deformation characteristics of the finite elements constituting the structural behavior are usually assumed to be deterministic. This is not to say that it is not possible to model system properties by random vectors so that even if the loading is deterministic the load effects will be uncertain. In fact a theory of stochastic finite element has been and still is being developed. The basic formulations and techniques are already available. It is only the extremely time consuming computation which at present prohibits the practical use of this methodology. It has also been shown in a number of numerical examples that the influence of uncertain

deformation properties in the system is usually small to negligible and thus would have to be taken into account only in extreme cases. The influence of the manner in which the finite elements are defined, in particular, their number and their spatial separation has to be checked as it should be clear that any discretization of an initially continuous problem necessarily leads to unconservative reliability estimates (artificial reduction of failure modes). Depending on the number of elements and their configurations implying system collapse the number of paths to system failure can be extremely large. It is therefore mandatory in practical analyses to limit the number of elements as far as possible.

Most results are available for linear—elastic and for rigid plastic frame or truss structures and to some degree also for linear elastic—plastic behavior of the components. The classical failure criterion of a redundant, not perfectly rigid plastic system is the singularity of the stiffness matrix. This singularity can be reached in a number of sequences of structural and componental state changes. Thus, the different singularity events form a series system. Intermediate system states during system degradation by componental failures (state changes) can be modeled by the union of cut sets of Boolean failure events. More recent attempts to refine the mechanical model primarily concentrate on modifications of the linear elastic—plastic model, e.g. by introducing semi—brittleness (reduced plastic domain beyond elasticity) or by introducing bilinearity (see, for example, Melchers/Tang, 1984; Rashedi/Moses, 1988). This increases the number of componental states. Depending on the number of components and of different states of the components the system can have an enormous number of system states. In applications this significantly can limit the degree of detail into which a modeling of componental behavior is feasible.

The earlier attempts to account for the usually many paths to failure in an approximate way are based on the so called incremental load method (see, for example, Moses, 1982; Moses/Stahl, 1978; Moses/Rashedi, 1983). At each load increment along a prescribed load path the state changes in the components are recorded. The mechanical model is adjusted and the most likely or some additional less likely failure modes are determined. By far the most researchers, however, used the similar so called failure tree approach for the analysis of redundant structural systems starting with the work of Murotsu (1981). This type of approach has been accustomed from well known approaches in classical reliability. It rests on the identification and analysis of the sequence of componental state changes (failures) from the initial intact state of the structure to at least the dominating failure modes of the system. In particular, if in a certain failure branch i failures have already taken place the mechanics for the i+1—th failure have to take account of the i previous componental state changes and the i+1—th failure event can be written as

$$F^{i+1} = F^{1} \cap F^{2|1} \cap F^{3|1\cap 2} \cap \ldots \cap F^{i|1\cap 2\cap \ldots \cap i-1} \cap F^{i+1|1\cap 2\cap \ldots \cap i} \tag{12}$$

Note that in failure events on the right hand side of this equation the changes in the mechanical behavior of the system introduced by the failures left to the considered event have to be taken into account appropriately.

The probability of eq. (12) can well be computed by FORM/SORM. Mechanically, provisions must also be taken that the sequence of state changes is a valid sequence which could occur along the possible load paths. Murotsu (1981) has introduced the concept of branch and bound. Its basic idea is to concentrate on the analysis of the failure branch with largest failure probability. Murotsu simplified the analysis in a number of ways. For example, he and others proposed to determine the most likely next component to fail by intersecting the failure domain of the first (or the temporarily last) failure domain with the failure domain of all remaining intact components. The book by Thoft—Christensen/Murotsu (1986) contains many details about the approach. Guenard (1984), Grimmelt (1984) and Gollwitzer (1986) improved the scheme in that they not only computed the full intersection along a failure path by FORM/SORM but continued the algorithm at a branching point in any of the already investigated failure branches

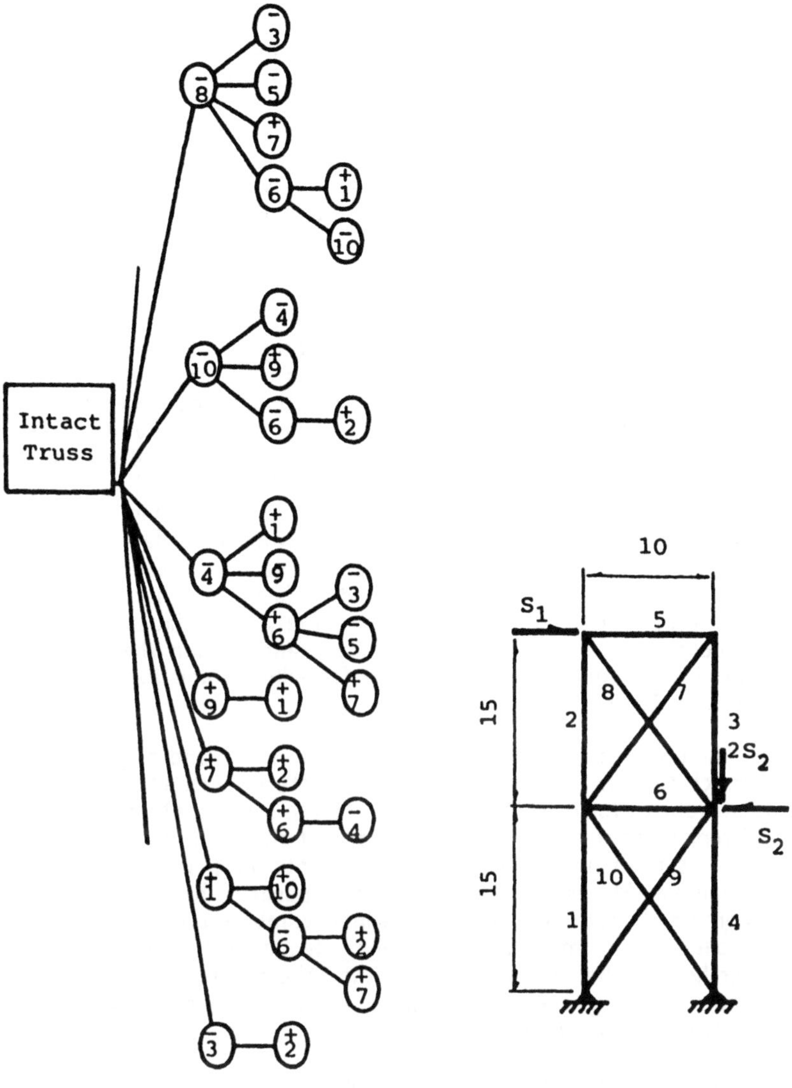

Fig.: 11: Failure Tree for Truss Structure (after Moses, 1990)

where the intersection probability is momentarily largest. This algorithm leads to a number of completed failure paths which should include the most dangerous failure path and a generally much larger number of uncompleted failure paths (see fig. 11). The probability of the union of all completed failure paths is a lower bound to the system failure probability and the union of all completed and uncompleted failure paths provides an upper bound to the failure probability. If the analysis is terminated before a complete failure path is found the probability of the union of all investigated uncompleted failure paths is an upper bound to the system failure probability which in general is much smaller than the simple series system bound. Quite frequently it suffices to investigate only rather short incomplete failure branches in practical applications.

It needs not to be emphasized that a rigorous pursuit of this approach can be an extremely laborious especially for larger and well designed structural systems. The analysis of some not completely unusual structures has also shown that care must be taken to introduce simplifications such as the one mentioned above because then there is a large likelihood to miss the important branches. On the other hand the failure tree approach is very flexible to account, for example, for P–δ effects and overall instability (Casciati, 1978; Casciati/Faravelli, 1980; Gollwitzer/Rackwitz, 1985), yielding in some components as well as unloading in certain components provided that the loading path is prespecified. Overall instability is simply taken as an additional failure event at each branching point. Yielding of a component can be introduced either on the mechanical side by introducing a yield surface or a set of yield surfaces together with the relevant flow rules or on the probabilistic side by introducing the yield surfaces as equalities into the intersection operation for the events along the branch. Then the path probability has to be computed as a conditional probability for which some tools have also been developed (Schall et al. 1988).

More recently the failure tree approach together with a branch and bound algorithm has been developed much further for special types of structures such as frames as they are used as support structures for offshore platforms. In the same context the possibility of a at least partial replacement of the branch and bound algorithm by certain importance sampling schemes has been investigated. Such attempts have, in fact, be made much earlier (see, for example, Moses/Fu, 1988) and still appear to be an interesting area of research. Such combined methods, i.e. those which combine intelligently the rigorous failure tree approach with importance sampling techniques may be called hybrid approaches. Nevertheless, the studies indicate that the failure tree approach with or without importance sampling even in simplified form is only practicable for not too highly redundant structures at present.

The most serious type of idealization for non–rigid–plastic models at present nevertheless is a loading either on prescribed deterministic paths or simply by proportional loading. In practical applications there is a need to first "load" the structure by its uncertain dead load and expose it to a combination of random environmental and operational loads. For offshore structures it is desirable to compute system degradation under a full sinusoidal extreme wave load including load–effect reversals below yielding forces. In principle, it is possible to take account of random, non–proportional load paths within the failure tree approach (see, for example, Karamanchandani, 1990). It must, however, be stated that, then, extremely large failure trees need to be investigated. Thus this method is restricted to research applications.

In view of the difficulties with the failure tree approach which nevertheless is the only approach capable to take account of realistic mechanical behavior it appears necessary to reconsider the basic concepts of system reliability analysis. There is no doubt that one of the obstacles for an efficient analysis is the Boolean description of componental and system states. At present two alternatives seem to be under study. In the first the singularity criterion of the stiffness matrix is replaced by a global deformation criterion for system states in the stable domain. Such a criterion must be conservative and together with an appropriate mechanical formulation is computationally feasible even for large structures. The degree of conservatism and the way in which such a global deformation criterion has to be formulated is not yet clear. The second alternative consists in replacing the Boolean description of componental states by a smooth function. This function increases smoothly from zero to one as component degradation progresses. With this choice also the system states are described by a smooth, yet not necessarily monotonic function. A suitable chosen envelope, however, can be made "sufficiently" smooth and the system failure probability can be determined by an expectation operation over the domain of uncertain variables.

Structures with a rigid plastic behavior of its components have been studied by a large number of investigators and a fairly complete set of results is available based both on the lower and upper bound theorems of plasticity theory (Augusti/Baratta, 1972, Ditlevsen/Bjerager, 1984). This is partly due to the fact that the question of load paths and thus the sequence of componental failures is irrelevant for the final collapse. Depending on the stochastic structure of the resistances (yield forces) in the components it was found that such structures can have a significant extra reliability due to redundancy. It appears, however, that if such structures were optimized with respect to reliability the beneficial effect of redundancy vanishes to a large extent. Although the rigid plasticity model is a valid mechanical model only for a few types of material and structural layouts it is a limiting case for materials which behave elastic–plastic or even non–linear. Recently successful attempts have been undertaken to take account of interaction effects in the components via so–called flow rules at various levels of sophistication indicating that substantially different reliabilities can be determined as compared to approaches where flow rules are not taken into account. Earlier approaches without inclusion of specific flow rules must, therefore, be considered as crude approximations. As Ditlevsen/Bjerager (1986) have pointed out the lower bound theorem of plasticity is only valid if there exists a statically admissible set of internal forces in all the potential hinges (yield zones) such that these internal forces nowhere violate the (strain–independent, convex) yield condition under observation of the relevant flow rule. For the much more used upper bound theorem of plasticity a similar restriction holds. The upper bound theorem says that if a kinematically admissible set of rates of deformations (strains) imposed at the yield hinges under the validity of the given flow rule exists and the corresponding plastic dissipation is at most equal to the rate of work done by the external forces then the structure can no more carry the load. While upper bound solutions have been investigated successfully in recent years in many papers (see Ditlevsen/Bjerager, 1986, for a review), the lower bound solutions still are relatively few and require considerable computational efforts. It is neither possible nor necessary here to elaborate on the details of the approaches put forward so far. They distinguish themselves more by their techniques rather than by their concepts. In principle, in the lower bound approaches the task consists in defining the safe set of structural states by making use of as much redundancy as is present in the structure. Usually not all redundancy degrees are exhausted and therefore the lower bound approach also yields a lower bound from the reliability point of view. It is important to know that the identification of "critical safe sets" is not necessary although their knowledge can improve the reliability bound substantially. The lower bound approach must identify the failure set of structural states and here it is necessary to identify the critical set providing an upper bound to system reliability. Again we refer to Ditlevsen/Bjerager (1986) for more details.

The other extreme of perfect elastic–brittle components in structures of general topology has most thoroughly been investigated by Bjerager (1985) and Bennet (1985). Mostly, the failure tree approach is applied. Formally, there are no specific methodological aspects except that there is an increased likelihood of successive componental failures until a new equilibrium state is reached under the load responsible for the componental failure initiating that succession of failures. From an engineering point of view it is interesting that in a number of relevant studies it was found that the assumption of brittleness can diminish the effect of redundancy substantially. Extremists will even say that there is no effective redundancy. Moreover, such systems behave like series systems which under otherwise consistent conditions are less reliable than the simple component. Those observations are consistent with the findings for the brittle Daniels system.

It is necessary to mention yet another class of methods which is capable to handle arbitrary structures and which is found increasingly in practical applications but less in research. The so called resonse surface approach attempts to avoid the difficulties in the probabilistic calculations with domains given by unions of intersections by appximating the boundary between safe system states and the failure states by simple linear to quadratic interpolation surfaces or even regression surfaces between preselected points in the basic variable space. Each of the points can be calculated mechanically as accurate as possible. Also, the stochastic model can be arbitrarily sophisticated. The accuracy of the method, of course, depends on the location of the selected points and the ability of the response surface to maintain the probability content of the real failure domain. This method is only approximate and in general it is not easy to quantify the error. At present it must be considered as a last resort when other methods fail to produce a solution.

6. Summary and conclusions

It is not easy to summarize the results in the studies with respect to general insights but dominating factors have been clearly identified. Those are

- the ratio of the load variability against the variability of system resistance,
- the topological arrangement of structural members,
- the mechanical behavior of the components and
- the dependence structure of the resisting variables.

Positive dependence (positive correlation) between the resistances in a series system improves its reliability whereas it can dramatically diminish the reliability in redundant systems. Large load variability makes any system consideration less important because redundant components provide only little extra reliability. If the first component in such system fails it is simply rather likely that the redundant elements also fail due to the large variability in the load and the internal load–effect redistribution. If the topological arrangement in the system is such that the additional load after failure of a component in the course of load redistribution affects primarily the adjacent elements little effect on reliability can in general be gained by extra redundant components. Then, load–effect redistribution alone can lead to progressive collapse. Finally, the degree of ductility of the members appears to be extremely important. If it is not really large the variability of the ductility limit also enters as an important factor. All in all, the general findings for the ideal Daniels system are confirmed also for systems of common topology.

These observations limit the practical significance of structural redundancy with respect to extra reliability. If one remembers that realistic structural systems tend to fall into the category of systems with distinct local load–sharing and which, therefore, behave almost like weakest–link structures one is lead to the conclusion that use of the extra reliability of redundant structures must be made with utmost care. The factors in favor for extra reliability by redundancy such as independence of componental resistances and ductility need to be verified. In particular, the assumption of full plastic componental behavior made in many structural system reliability studies but also in practical design may lead to gross overestimations of system reliability.

The foregoing observations have triggered many discussions about the definition of redundancy and robustness and the possible ways of their quantification. At this place it appears appropriate to recall the definitions proposed in the conclusions of the NSF–workshop on "New Directions in Structural System Reliability" (see the proceedings in Struct. Saf., Vol. 7, No. 2–4, 1990). Redundancy simply is analogous to static indeterminacy where it has been well recognized that for continuous structures this definition must be related to the manner in which it is subdivided into components. But no general agreement on the definition of a quantitative reliability–based measure of redundancy could be reached. Many of the specialists favored a measure by the ratio of the system failure probability to the failure probability of any component. Since structures can have rather weak components whose failure affects system reliability very little the weakest or the weaker components are not considered as appropriate references. Robustness of a system, on the other hand, has been defined as the ability to remain functional after failure of one or more of its components. It can be measured by the conditional system failure probability the condition being the failure of some of its components.

Because many counteracting factors affect time–invariant structural system reliability, much further research is needed before it is justified to overrule the reluctance in direct practical applications and in code making to generally apply the findings and methods in system reliability. It is hoped that the foregoing considerations made it clear that any inadequate simplification can lead to erroneous and misleading results. Realistic stochastic modeling and efficient and accurate reliability methods are required as well as sound mechanical modeling. Further fundamental research is necessary to improve and to speed up the reliability calculations. Simultaneously, and probably more important, research towards more realistic mechanical models compatible with the probabilistic part is needed. In particular, if the extra reliability due to redundancy is really to be exhausted, then, structural system behavior close to its collapse and, at the compnent level, beyond the bearing capacity, needs to modeled appropriately. Finally, the area of time–variant system reliability is still an open field with promise of substantial progress in concepts and methods.

REFERENCES

Augusti, G., Baratta, A., Limit Analysis of Structures with Stochastic Strength Variations, J. Struct. Mech., 1, 1, 1972, pp. 43—62

Barbour, A.D., Brownian Motion and a Sharply Curved Boundary, Adv. Appl. Prob., 13, 1981, pp. 736—750.

Bennett, R.M., Reliability Analysis of Frame Structures with Brittle Components, Structural Safety, 2, 1985, pp. 281—290

Bjerager, P., On Computation Methods for Structural Reliability Analysis, Struct. Saf., 9, 2, 1990, pp. 79—96

Bjerager, P., Reliability of Brittle Structural Systems, Proc. ICOSSAR'85, Vol. 1, 1985, pp. 149—155

Breitung, K., Asymptotic Approximations for Multinormal Integrals, Journ. of the Eng. Mech. Div., Vol. 110, No.3, 1984, pp. 357—366

Breitung, K., Hohenbichler, M., Asymptotic Approximations for Multivariate Integrals with an Application to Multinormal Probabilities, Journ. of Multivariate Analysis, Vol. 30, No.1, 1989, pp. 80—97

Casciati, F., A Probabilistic Approach to the Deformation Analysis of Elastic—Plastic Frames, J. Struct. Mech., 6, 1, 1978

Casciati, F.,Faravelli, L., Safety Analysis for Random Elastic Plastic Frames in the Presence of Second Order Geometrical Effects, Appl. Math. Mod., 1, 4, 1980

Daniels, H.E., The Statistical Theory of the Strength of Bundles of Threads, Part I, Proc. Roy. Soc., A 183 (1945) pp. 405—435.

Daniels, H. E., The Maximum Size of a Closed Epidemic, Adv. Appl. Prob., 6 (1974) pp. 607—621.

Daniels, H.E., The Maximum of a Gaussian Process whose Mean Path has a Maximum, with an Application to the Strength of Bundles Fibres, Adv. Appl. Prob., 21, 1989, pp. 315—333

Ditlevsen, O., Narrow Reliability Bounds for Structural Systems, Journ. of Struct. Mech., Vol. 7, No. 4, 1979, pp. 453—472

Ditlevsen, O., Gaussian Outcrossings from Safe Convex Polyhedrons, Journ. of the Eng. Mech. Div., ASCE, Vol. 109, 1983, pp. 127—148

Ditlevsen, O., Bjerager, P., Reliability of Highly Redundant Plastic Structures, J. Eng. Mech., ASCE, 110, (5), 1984, pp. 671—693

Ditlevsen, O., Bjerager, P., Methods of Structural Systems Reliability, Struct. Saf., 3, 3+4, pp. 195—229, 1986

Fujita, M., Grigoriu, M., Rackwitz, R., Reliability of Daniels—Systems Oscillators Including Dynamic Redistribution, Proc. 5th ASCE Specialty Conference "Probabilistic Methods in Civil Engineering", Blacksburg, Va., 1988, pp. 424—427

Giannini, R., Nuti, C., Pinto, P.E., Reliability Analysis of Nuclear Systems under Seismic Excitation, Proc. ICOSSAR '85, Kobe, (Ed. I. Konishi et al.) IASSAR, Vol.III, 1985, pp. 223—231

Gollwitzer, S., Zuverlässigkeit redundanter Tragsysteme bei geometrischer und stofflicher Nichtlinearität, Dissertation, München, 1986.

Gollwitzer, S., Rackwitz, R., First—Order Reliability of Structural Systems, Proc. ICOSSAR'85, Vol. 1, 1985, pp. 171—180

Gollwitzer, S., Rackwitz, R., An Efficient Numerical Solution to the Multinormal Integral, Prob. Eng. Mech., 3, 2, 1988, pp. 98—101

Gollwitzer, S., Rackwitz, R., On the Reliability of Daniels Systems, Struct. Saf., 7, 2—4, 1990, pp. 229—243

Grigoriu, M., Reliability of Degrading Dynamic Systems, Struct. Safety, 8, 1990, pp. 345—351

Grimmelt, M., Eine Methode zur Berechnung der Zuverlässigkeit von Tragsystemen unter kombinierten Belastungen, Diss., Berichte zur Zuverlässigkeitstheorie der Bauwerke, SFB 96, Technische Universität München, Heft 76, 1984

Gücer, D.E., Gurland, J., Comparison of the Statistics of two Fracture Moments, J. Mech. Phys. Solids, 10, 1962, pp. 363—373

Guenard, Y.F., Application of System Reliability Analysis to Offshore Structures, John A. Blume Engineering Center, Thesis, Report No. 71, Stanford University, 1984

Guers, F., Rackwitz R., Time—Variant Reliability of Structural Systems Subject to Fatigue, Proc. ICASP 5, Vancouver, Vol. I, 1987, pp. 497—505.

Guers, F., Dolinski K., Rackwitz R., Probability of Failure of Brittle Redundant Structural Systems in Time, Structural Safety, 5, 1988, pp. 169—185.

Harlow, D. G., Phoenix S. L., Probability Distributions for the Strength of Composite Materials, Int. Journ. Fracture, 17, 4, 1981

Henley, E.J., Kumamoto, H., Reliability Engineering and Risk Assessment, Prentice—Hall, Englewood Cliffs, New York, 1981

Hohenbichler, M., Resistance of Large Brittle Parallel Systems, Proc. 4th ICASP—Cont., Universita di Firence, Italy, 1983, pp. 13O1—1312.

Hohenbichler, M., Rackwitz, R., On Structural Reliability of Brittle Parallel Systems. Reliability Engineering, 2, 1981, pp. 1—6.

Hohenbichler, M., Rackwitz, R., Reliability of Parallel Systems under Imposed Uniform Strain, Journ. Eng. Mech. Div., ASCE, 109, 3 (1983) pp.896—907.

Hohenbichler, M., Gollwitzer, S., Rackwitz, R., Parallel Structural Systems with Non—linear Stress—strain Behavior, SFB 96, Technische Universität München, 58, 1981, pp. 23—54

Hohenbichler, M., Rackwitz, R., First—Order Concepts in System Reliability, Structural Safety, 1, 3, 1983, pp. 177—188.

Hohenbichler, M., Rackwitz, R., Asymptotic Crossing Rate of Gaussian Vector Processes into Intersections of Failure Domains, Probabilistic Engineering Mechanics, Vol.1, No.3, 1986, pp. 177—179

Hohenbichler, M., Gollwitzer, S., Kruse, W., Rackwitz, R., New Light on First— and Second—Order Reliability Methods, Structural Safety, 4 (1987) pp. 267—284.

Leadbetter, M.R., Lindgren, G., Rootzen, H., Extremes and Related Properties of Random Sequences and Processes, Springer, New York, 1983

Karamandchani, A., New Methods in System Reliability, Ph..D. Thesis, Dept. of Civil Eng., Stanford University, Stanford, 1990

Jorgensen, J.L., Goldberg, J.E., Probability of Plastic Collapse failure, J. Struct. Div., ASCE 95 (ST8), 1969, pp. 1743—1761

Kersken—Bradley M., Beanspruchbarkeit von Bauteilquerschnitten bei streuenden Kenngrößen des Kraftverformungsverhaltens innerhalb des Querschnittes, Zuverlässigkeitstheorie der Bauwerke, SFB 96, Technische Universität München, 56, 1981,

Melchers, R.E., Tang L.K., Dominant Failure Modes in Stochastic Structural Systems, Structural Safety, 2, 1984, pp. 127—143

Moses, F., System Reliability Developments in Structural Engineering, Structural Safety, Vol.1, No.1, 1982, pp. 3—13

Moses, F., Kinser, D.E., Optimum Structural Design with Failure Probability Constraints, AIAA Journ., Vol. 5, 6, 1967, pp. 1152—1158

Moses, F., Stahl, B., Reliability Analysis Format for Offshore Structures, Proc. Offshore Techn. Conf., Houston, 1978

Moses, F., Rashedi, M. R., The Application of System Reliability to Structural Safety, Proc. ICASP—4, Florence, 1983, pp. 573—584

Moses, F., Fu, G., Importance Sampling in Structural System Reliability, ASCE Spec. Conf. on Probabilistic Mechanics, Blacksgurgh, 1988

Moses, F., New Directions and Research Needs in System Reliability Research, Struct. Saf., 7, 2—4, 1990, pp. 93—100

Murotsu, Y., Okada, H., Yonezawa, M, Taguchi, K., Reliability Assessment of Redundant Structure, Proc. ICOSSAR'81, Structural Safety and Reliability, Trondheim, Elsevier, Amsterdam, 1981, pp. 315—329

Phoenix, L.S., Taylor, H.M., The Asymptotic Strength Distribution of a General Fiber Bundle, Adv. Appl. Prob., 5, 1973, pp. 2OO—216

Rackwitz, R., Reliability of Systems under Renewal Pulse Loading, Journ. of Eng. Mech., ASCE, Vol. 111, No. 9, 1985, pp. 1175—1184

Ramachandran, K., Baker, M., New Reliability Bound for Series Systems, Proc. ICOSSAR '85, Kobe, (Eds. I.Konishi et al.), IASSAR, Vol.I,1985, pp. 157—169

Rashedi, M.R., Moses, F., Identification of failure modes in system reliability, ASCE, J. Struct. Eng., 114, 1988, pp. 292—313

Rosen, D.W., Tensile Failure of Fibrous Composites, AIAA Journal, 2, 1964, pp. 1985—1991

Sen, P.K., Bhattacharyya, B.B., Asymptotic Normality of the Extremum of Certain Sample Functions, Zeitschrift zur Wahrscheinlichkeitstheorie und verwandte Gebiete, Vol. 34, 1976, pp. 113—118

Schall, G., Gollwitzer, S., Rackwitz, R., Integration of Multinormal Densities on Surfaces, Proc. 2nd WG 7.5 Work. Conf., Springer, 1988

Schrupp, K., Rackwitz, R., Outcrossing Rates of Gaussian Vector Processes for Outsets of Componential Failure Domains, in: Proc. ICOSSAR'85, Kobe, (Ed. I. Konishi et al.), IASSAR, 1985, pp. 6O1—6O9

Shinozuka, M., Itagaki, H., On the Reliability of Redundant Structures, Ann. Rel. Maint., SAE—ASME—AIAA, 5 (1966).

Smith, R.L., Limit Theorems and Approximations for the Reliability of Load—sharing Systems, Adv. Appl. Prob., 15, 1983, pp. 304—330.

Smith, R.L., Phoenix, S.L., Asymptotic Distributions for the Failure of Fibrious Materials under Series—Parallel Structure and Equal Load—Sharing, J. Appl. Mech., 48, 1981, pp. 75—82.

Stevenson, J., Moses, F., Reliability Analysis of Frame Structures, J. Struct.Div., ASCE, 96 (ST11), 1970, pp. 2409—2427

Thoft—Christensen, P., Murotsu, Y., Application of Structural Systems Reliability Theory, Springer, Berlin, 1986

Vanmarcke, E.H., Matrix Formulation for Reliability Analysis and Reliability—Based Design, Computers & Structures, 3, 1973, pp. 757—770

Wen, Y.K., Chen, H.—C., System Reliability under Multiple Hazards, Civil Engineering Studies, Structural Research Series No. 526, Report No. 2OO7, Department of Civil Engineering, University of Illinois at Urbana, Illinois, 1986

Yao, T. P., Yeh, H.—Y., Formulation of Structural Reliability, J. Struct. Div., ASCE, 95, ST12, 1968, pp. 1—15

Zweben, C., Tensile Failure Analysis of Fibrous Composites, AIAA Journal, 6, 1968, pp. 2325—2331

MATERIAL CHARACTERISTICS AND RELIABILITY-BASED DESIGN

R. O. FOSCHI
Department of Civil Engineering
University of British Columbia
Vancouver, Canada V6T 1Z4

ABSTRACT. Characteristic material properties and associated topics, relevant to the development of reliability-based design procedures for wood, are discussed. The focus is on issues concerning short and long-term mechanical behavior, size effects and environmental influences.

1. Introduction

The objective of reliability–based design is the systematic consideration of all the uncertainties involved in the design process, in such a way that the probability of the structure not performing as intended can be quantified. The uncertainties may refer to the description of the applied loads, the estimation of material strength or the model used to calculate the effect of the loads. Principles of reliability–based design may be used in two ways: either to estimate the reliability of a given structure under load or to calibrate simple design equations for codes. These equations attempt to provide adequate reliability levels, specified a–priori, for all possible cases to which the design equation may apply. To this end, the applicability of the equation is enhanced by the calibration of several factors, some attached to the load effects and some to the strength. The resulting design procedure has been given different names: for example, LRFD (Load and Resistance Factor Design) in the United States, LSD (Limit States Design) in Canada. In general, these equations are of the form

$$\alpha_D D_n S_D + \alpha_Q Q_n S_Q = \phi R_n K \tag{1}$$

where D_n, Q_n, α_D and α_Q represent, respectively, the "design" dead and live loads and their corresponding "load factors"; R_n is a "characteristic" strength; ϕ is the "resistance factor" or "performance factor"; and K represents a variety of strength modifications relevant to the material behavior and structural application of the member. S_D and S_Q are factors (deterministic or random) to transform applied loads into capacities or stresses.

Codes in North America have specified load factors which are the same for all materials. This has meant that calibration of design guidelines for wood, for example, coming after those for steel, had to use pre–existing load factors and had only the freedom to calibrate those related to the material strength.

J. Bodig (ed.), Reliability-Based Design of Engineered Wood Structures, 75–89.

When the problem is one of assessing the reliability of a particular application, a performance function G is written such that

$$G = R - (DS_D + QS_Q) \tag{2}$$

where D and Q represent, respectively, the dead and live random load variables, and R is the random strength variable consistent with the application. Thus, R may be the capacity modified according to the environmental conditions under which reliability is being estimated. Probability of non-performance corresponds then to the probability of $G < 0$, which can be estimated from the calculation of the reliability index β by well established algorithms.

In code calibration, Eq.(1) is combined with Eq.(2) to give

$$G = R - \frac{\phi R_n K}{\alpha_D \gamma S_D + \alpha_Q S_Q} [\gamma S_D d + S_Q q] \tag{3}$$

where $\gamma = D_n/Q_n$ is the ratio between design dead and live loads, $d = D/D_n$, and $q = Q/Q_n$. Using Eq.(3), the reliability index can be estimated in terms of the resistance factor, ϕ, given the load factors and a ratio γ. An assigned target reliability level can then be used to find the corresponding design factor, ϕ. Note that R must be modified as required by the end use: for example, for bending under high moisture conditions, R must be the distribution for this particular situation. However, the selection of a characteristic strength, R_n, or the "design adjustment", K, is to some extent, arbitrary. Of course, K must represent the actual material behavior, but there is here some freedom to use a simplified modification factor if the actual material model is too complicated for code purposes.

The objective of this paper is to discuss some of the most important aspects of wood material behavior as they affect the implementation of reliability–based design. Although there are now many composite wood products which may be classified as "manufactured" with relatively controlled properties, this paper will mainly cite examples from lumber. However, the topics covered apply equally to reliability–based design for the more advanced composite products.

2. The Material

The statistical characterization of the strength, R, is not a simple proposition. For a particular application, R could be determined directly from tests reproducing the actual service conditions. Although always an alternative, this is not the desirable or efficient approach. On the other hand, tests can be used to provide information on R under certain conditions, and the results extrapolated to other situations with the utilization of behavior models. Such models must be verified, and when operative they permit the quantification of R required for reliability studies. Of course, not only strength must be predicted, but also stiffness which is required for serviceability limit states.

Many applications still use wood as originally obtained from the tree. In its natural state, wood is highly variable, and mechanically complicated. As the tree grows forming rings, approximately circular, of different densities, the closest mechanical material model would be that of a body with cylindrical anisotropy, with the axis of the cylinder in the direction of the wood fibers. However, this body would have to be nonhomogeneous and the fiber direction itself would have to change along the axis. Furthermore, the normally rectangular boundaries would mathematically conflict with the cylindrical description of the material. A model of such complexity would result in problems always requiring numerical solutions, and one of the most obvious difficulties would be the gathering of the appropriate material data. Normally, problems are simplified by considering a plane orthotropic model, considering only two directions: parallel or perpendicular to the fibers.

Moduli of elasticity, shear moduli and Poisson ratios are defined according to the plane orthotropic model and assumed to obey the appropriate symmetries. Normally, material behavior is studied along the two predominant directions: thus, data are obtained for tension parallel to the fibers (or "grain"), or for the perpendicular direction. Although, in reality, the material will be subjected to two or three–dimensional stress states, behavior under such conditions has not been studied to the same degree. Wood may be considered as a bundle of fibers in a matrix, and its strength limits may be reached by either failing of the fibers or by breaking up the matrix bond. According to one of the models, parallel capacity is controlled by fiber strength, while perpendicular and shear capacity is related more to the strength of the matrix. The model is normally simplified even further by assuming that the material properties do not change from point to point. For example, the modulus of elasticity E used in calculations is normally an "average" obtained from the stiffness of a specimen of structural dimensions. Nevertheless, it is well known that stiffness properties are nonhomogeneous and that, for example, E varies substantially along the length of a specimen. Similarly, the tensile strength of the fibers varies along their direction, and a tension test only measures the minimum strength within the specific length of the specimen. Thus, nonhomogeneous properties are characteristics of wood which influence reliability, and are the cause of size effects which must be assessed in specific applications.

Strength and stiffness are also influenced by the history of the applied load. For strength, this phenomenon is known as the "load duration" effect. In essence, it is not different from the fatigue phenomenon in metals, except that a wood member may fail over time even under a constant load. The load duration behavior of the material, including both the long and very short–term, must be evaluated and quantified for the prediction of reliability within the intended service life of the structure. The effect of time on stiffness, or creep/relaxation phenomenon, must be similarly studied for the evaluation of both serviceability and strength limit states. In addition, there is an interaction between the two phenomena: a material which exhibits accelerated creep also tends to show more duration of load effect. Thus, increasing the moisture content of a specimen, which increases its creep deformation, may shorten its life under constant load. The interactions between long–term strength, creep, and moisture/temperature states are thus important and must also be quantified for reliability assessment.

Since changes in moisture content or temperature, resulting from environmental variations, do not occur uniformly throughout the volume of a specimen, cracks tend to appear and propagate. These cracks directly influence the capacity of the fiber–bonding matrix, and thus affect the capacity in shear and tension perpendicular to the fibers. Cracks may also be introduced when machining for mechanical fasteners, or driving nails or dowels. Reliability assessment, when appropriate, must therefore consider the presence of cracks, requiring an understanding and quantification of crack growth mechanics in wood.

Mechanical properties may also be influenced by treatments to prevent decay or to increase resistance to fire, with a corresponding effect on reliability.

Finally, but not less important, the structural reliability of wood products is influenced by their classification (grading) and the different quality controls implemented. Product strength variability can be controlled by non–destructive evaluation of strength indicators (e.g. stiffness, grain direction, density), or directly by proof–testing. Reliability assessment requires statistical data for each grade or classification, and the estimate will be influenced by the sampling used in the data gathering. Thus, statistics for a given size, species and grade of lumber, if they are intended to reflect the entire production, must be appropriately applied to a weighted sample representing different producers and growing regions. The uncertainty in the derived statistical parameters must also be included in the reliability assessment. For each of the material characteristics just mentioned, specific problems must be addressed and these will now be discussed in more detail.

3. Short–Term Strength

Basic strength information is obtained under a ramp load history of short duration (1 to 5 minutes on the average). The distribution of values thus obtained is used to determine the characteristic strength for the design equations. In the traditional working stress design procedures, the allowable stress was equal to the 5th–percentile of the short–term strength distribution divided by a factor accounting for safety and duration of load. In the calibration of the Canadian reliability–based code (Foschi et al., 1989) the short–term 5th–percentile was again used as the characteristic strength R_n, having been obtained in Canada for each lumber species, cross–sectional size and grade, from a production–weighted sample of about 400 specimens per cell. The testing was done in bending, tension and compression parallel to grain, using specimens of structural length at an equilibrium moisture content of 15% . Because specimens were tested "in–grade" and in structural sizes, the data were more trustworthy, from a reliability viewpoint, than if they had been collected from small, clear specimens requiring further adjustment for size and influence of defects. Thus, it is important that short–term strength be evaluated using structural size specimens, as close to the actual end use applications as possible.

In Canada there were two such large scale in–grade testing programs. The first relied on proof–testing each specimen with a load greater than the estimated 5th–percentile of the distribution (Madsen, 1973), the second required the destruction of all 400 specimens in each data cell. For visually graded lumber, the coefficient of variation in bending strength can be as high as 0.45. The data for each cell need to be represented mathematically by a cumulative distribution function. The question arises as to which type of distribution is better to use. Although a normal distribution implies the possibility of negative values, it was considered along with a lognormal, a 2 and a 3–parameter Weibull distributions to represent the entire data cell. The corresponding relationship between the reliability index and the resistance factor was highly dependent on the type of distribution used. Although all four distribution types could be said to be good fits of the entire data set, they differed substantially on their representation of the lower tail, where the design points were always located. The problem can be eliminated by focussing on the lower tails (including data up to may be the 20th–percentile) if in doing so, of course, there are still a sufficient number

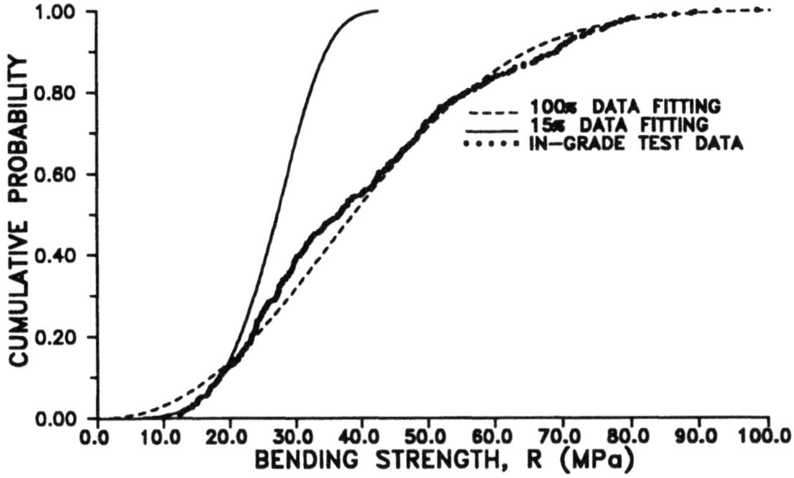

Figure 1. 2–P Weibull Distribution Fits to the Test Data (DF, No.2, 38mm × 184mm).

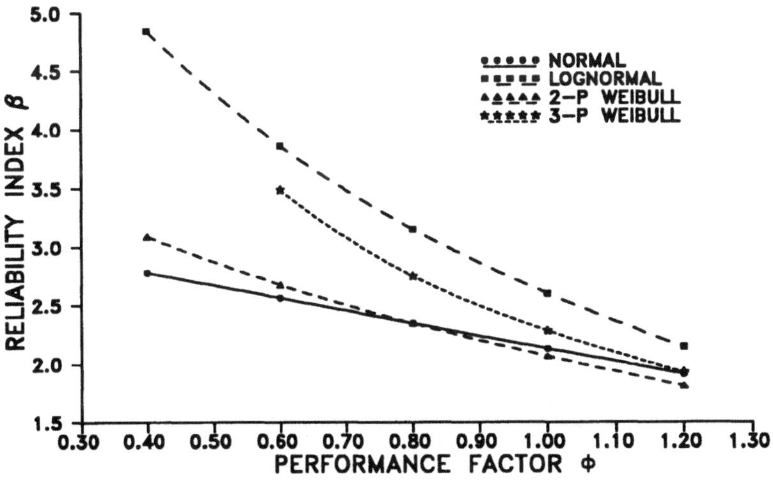

Figure 2. $\beta - \phi$ Relation for Four Distribution Types (SPF, 100% Data).

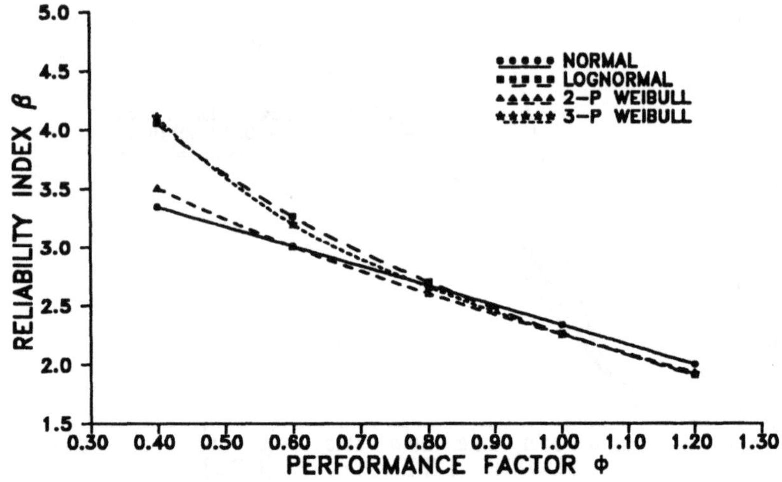

Figure 3. Relationship between β and ϕ for Four Distribution Types (SPF, Data Truncated at 15–th Percentile).

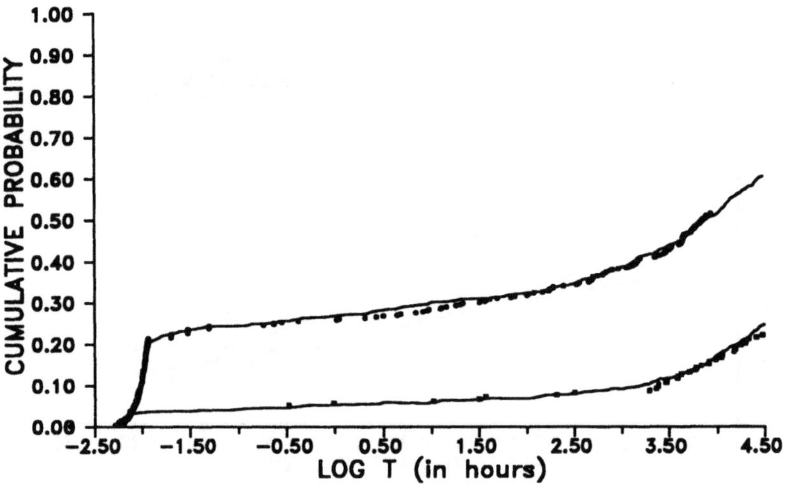

Figure 4. Canadian Model versus Experimental Results for Western Hemlock.

of data points included in the working set. Figure 1 shows short–term strength data from a bending test of Douglas fir (38 mm × 184 mm, No.2 grade) lumber, with a 2–parameter Weibull distribution fitted either to the entire set or to the lower 15th–percentile. These results are typical for other species, grades or sizes. Figure 2 shows the $\beta - \phi$ relationship for different distributions fitted to the entire set, while Figure 3 shows the results using distributions fitted to the lower 15th–percentile. Figures 2 and 3 correspond to No.2 grade spruce lumber, 38 mm × 184 mm, under 30 year maximum Quebec City snow load.

It is apparent that when good representation of the lower tail is achieved, the dependence on distribution type is very much diminished (at least at reliability levels of interest). The distribution fitting the lower tail does not represent the upper part of the data very well. Reliability results must ensure, nevertheless, that the design point always lies within the fitted range. This implies that testing programs based on proof loading could provide adequate information for reliability calibrations, at least as far as strength is concerned. For stiffness, since load sharing in systems is influenced by the entire range of moduli of elasticity, the entire distribution must be obtained.

Uncertainty in the strength distribution parameters can be studied by modifying Eq.(3) by replacing the random variable R with the corresponding distributional form. For the case of a 2-parameter Weibull distribution,

$$R = m[-\ln(1-p)]^{1/k} \tag{4}$$

The random variables are now the parameters m and k, and p (uniformly distributed between 0 and 1). The effect of replacing R with these additional variables is to reduce the reliability level. However, as shown by Foschi et al. (1989), the effect is not significant for sample sizes of the order of 400, resulting in coefficients of variation for m and k of the order of 0.05.

4. Size Effects

Size effects must be quantified in order to adjust test strength and stiffness for structural reliability analysis.

Because of nonhomogeneous spatial property distribution, short–term test data only apply to a population of specimens of the same geometry and size as that tested. Thus, if the population is of a grade which admits a range of defect sizes , the longer the specimen within the grade, the greater the chances that it may include the largest defect in the grade. This would tend to lower the average strength for the longer length and, at the same time, reduce the test variability. Perhaps because the range of possible sizes in specimen cross–section is much smaller than for the length, size effects are normally easier to detect from tests when the length of the specimen is varied. In theory, however, size dependence should be controlled by the volume of the specimen. Lumber test data in Canada and the U.S. have been obtained using specimens with a length to width ratio of 17, deemed to represent typical residential construction applications. For lack of a more fundamental fiber bundle fracture theory, size effects connected with failure parallel to grain have been represented using equations as if Weibull's theory of brittle fracture were applicable in the fiber direction. Essentially, for the case of tension, this implies that the strength, σ, of a specimen with dimensions B, H, L can be related to the strength of another specimen, σ_o, with dimensions B_o, H_o and L_o according to

$$\sigma = \sigma_o \left(\frac{BHL}{B_oH_oL_o}\right)^{1/k} \tag{5}$$

where k is the size effect exponent obtained from statistical regression of data. Equation(5) applies at any probability level, and has been used in Canada to adjust cell 5th–percentile values to smooth reliability results obtained by individual cell analyses (Foschi et al., 1989). Equation(5) can be used as a design size adjustment factor for the characteristic strength R_n, although it cannot be said that it provides a physical model for the size dependency.

On the other hand, size dependence in matrix strength (shear and tension perpendicular to grain) is more amenable to the assumptions of Weibull's brittle model. Accordingly, these size effects can be quantified using equations similar to Eq.(5). In the general case of a non–uniform stress field, the theory predicts a strength dependence on size as well as the distribution of stresses over the specimen volume.

Since the distribution of stiffness or modulus E is also nonhomogeneous, it must also result in size dependence for the overall stiffness of the specimen. However, it may be shown that the deflection of a beam is relatively insensitive to local variations in E, allowing use of the mean E value for the entire specimen. In some cases, however, local E variation may play a more important role: in glued–laminated beams, for example, the distribution of stresses between and along laminations is essentially controlled by the between and along lamination variations in E. Reliability studies in glued–laminated beams must therefore consider these variations, at least for strength limit states. Because of such applications, and because size effects in general are connected to spatial property variations, it is necessary to study these random processes in a more fundamental way. Although reliability studies can be conducted using approximate adjustments as shown in Eq.(5), important issues affecting non–destructive testing, quality control and reliability would surely benefit from a more fundamental study of these problems. A better understanding of the underlying processes may help reconcile results from experimental tests where the specimens have been positioned differently in the test span: sometimes with random location of the largest defect, sometimes with this defect purposely within the zone of largest stresses.

5. Duration of Load (DOL) and Creep

The phenomenon of duration of load links the strength of a wood specimen with the applied load history. In general, the specimen can support a higher load if this is of short duration, the strength degrading with time under a constant load. The duration of load effect was first studied using small, clear specimens in bending under constant load, and the results expressed in terms of applied stress ratio versus time–to–failure. Stress ratio was defined as the applied stress divided by the short–term strength of the specimen. Obviously, this is a quantity which cannot be determined by experimentation, and the original test used two closely matched specimens to approximate the ratio. The results are well known as the "Madison curve", and have been implemented in most codes. Although the data were restricted to the small specimens, the duration of load adjustments thus derived have been used for all types of applications. Tests with lumber pieces of structural size, under

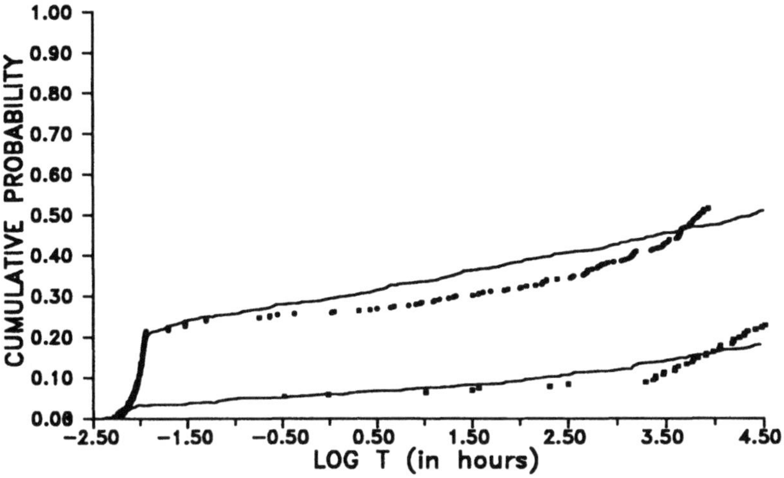

Figure 5. U.S. Model versus Experimental Results for Western Hemlock.

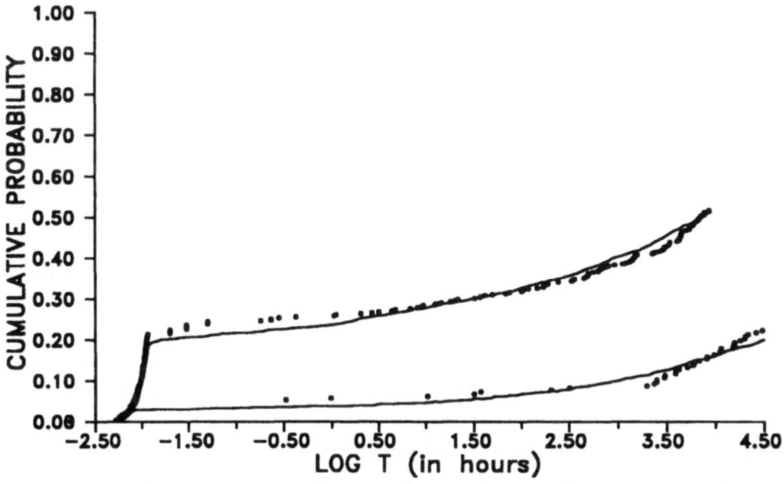

Figure 6. Nielsen's Model versus Experimental Results for Western Hemlock.

constant load for periods of up to 4 years, have since shown that the behavior of such pieces was substantially different (Madsen, 1976; Foschi and Barrett, 1982). Furthermore, tests at higher moisture contents have also shown a more pronounced duration of load effect.

Since tests are conducted with either a ramp or a constant load, means of extrapolating the results to intermittent service loads are required for the reliability analysis. This has been done using models for damage accumulation or for slow crack growth. A damage accumulation model has been used in Canada and is of the form

$$\frac{d\alpha}{dt} = a\,[\sigma(t) - \sigma_o\sigma_s]^b + c\,[\sigma(t) - \sigma_o\sigma_s]^n\,\alpha \tag{6}$$

where α is the damage parameter ($0 < \alpha < 1$), $\sigma(t)$ the applied stress history, and σ_s the short-term strength. Model parameters a, b, c, n and σ_o are calibrated to the results from constant load tests and σ_o is a threshold stress ratio, below which there is no accumulation of damage. Since the model must reproduce the short-term strength, σ_s, when the load history is the corresponding ramp load, only four model parameters are independent. It can be seen that the formulation in Eq.(6) assumes that damage growth is dependent on the stress level and the current damage. The latter, expressed by the second term in Eq.(6), introduces exponential growth and controls the process after some substantial damage has been accumulated.

Figure 4 shows constant load data from a Canadian experiment on hemlock, and the cdf of times-to-failure. The applied loads were the 5th- and the 20th-percentile of a control short-term strength distribution. Also shown in Figure 4 is the fit obtained with Eq.(6) when, in addition, the model parameters were assumed to be lognormally distributed between specimens. In particular, the best fit corresponded to a threshold stress ratio with a mean 0.53 and a coefficient of variation of 0.30. How important is the second term in Eq.(6)? If it is left out, the first term implies that damage growth is controlled only by the stress level. Under constant load, damage grows linearly and Figure 5 shows that such a model cannot represent the data trend very well. Although the overall fit is acceptable, failure rates are underestimated at longer failure times and overestimated up to about 6 months. A model equivalent to using only the first term in Eq.(6) has been adopted in the United States. The resulting duration of load adjustments are somewhat different from those obtained in Canada: reflecting the fit in Fig.5. Adjustments for permanent loads are less severe and the opposite occurs for service loads of snow for a life of 50 years.

A different model has been studied by Nielsen (1980), considering slow crack growth in a material with viscoelastic properties around the crack tip. This model utilizes a creep function with fluid characteristics and, accordingly, does not show a threshold. The advantage of this formulation is that it attempts to link creep with crack growth and damage. Since the speed of crack growth is proportional to the current crack length, the Nielsen model shows the same exponential growth as Eq.(6). In fact, Figure 6 shows how this model can fit the data equally well.

Load histories over the service life of the structure represent data from an stochastic load process. For each sample, the damage accumulation model permits the estimation of damage at the end of the service life. This quantity is random, and the probability of failure can be estimated by studying the performance function

$$G = 1.0 - \alpha(T) \tag{7}$$

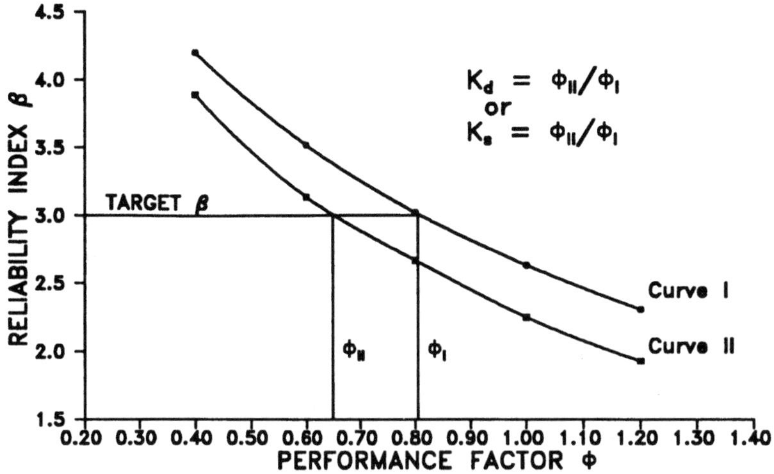

Figure 7. Procedure to Determine the DOL Adjustment Factor, K_d, or the Environmental Service Factor, K_s.

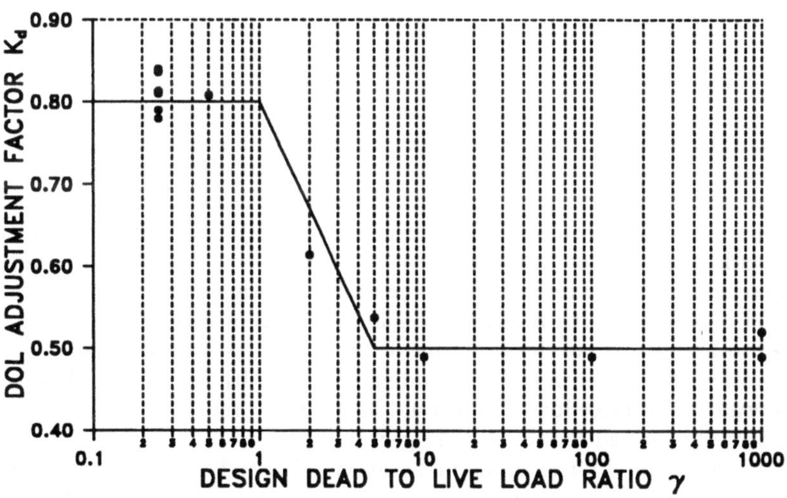

Figure 8. Recommended Relationship between γ and K_d.

where T is the service life. The probability of failure has been estimated using Monte Carlo simulations, and converted in the end to the corresponding reliability index, β.

It is important to stress that the duration of load adjustment, K_d, in the design equation cannot be derived directly from experimental data. Rather, it is part of the adjustment that must be made to the short–term resistance factor, when long term service loads are considered and the target reliability level must be maintained. Figure 7 shows the procedure. Curve I corresponds to the $\beta - \phi$ relationship for short–term loads, when the duration of load effect and damage accumulation is ignored. Here, the short–term loads are the maxima applied during the service life. The value ϕ_I corresponds to the target reliability . Curve II is obtained using Eq.(7) and includes damage accumulation. At the same target reliability, the factor must be ϕ_{II}. Obviously, if ϕ_I is used in the design equation, the factor K_d must be the ratio ϕ_{II}/ϕ_I.

Figure 8 shows the values of K_d introduced in the Canadian code. No large differences were observed between snow and occupancy loads, and a common value $K_d = 0.80$ was adopted. For constant loads over the service life, the value $K_d = 0.50$ must be used. Intermediate cases, for example snow loads with a high dead load component, must be interpolated between the two extremes. Further tests in Canada with two different qualities of spruce have shown no significant differences in K_d. However, since all the data correspond to dry lumber (in a semi–controlled covered environment), care must be taken when extrapolating them to wet service conditions. In this situation, increased creep may lead to a greater duration of load effect, and a corresponding more severe factor of K_d. Similarly, more severe K_d factors are to be expected, even for dry service conditions, when the material creeps more than solid wood, as it may occur for composite or reconstituted wood products. In the context of the damage accumulation model, a lower K_d factor results from a higher value of the parameter c or a lower threshold stress ratio σ_o.

Thus, duration of load and creep are phenomena which directly influence long–term reliability estimates. Although a relatively high confidence may be attached to estimates of K_d for dry lumber, more test results are required for other products with different creep characteristics. In general, a more complete understanding of the link between creep, moisture/temperature state and strength degradation is required (Hoffmeyer, 1990).

More research is needed on the effects of load duration in tension perpendicular to grain and shear, given their importance for the design of connections . In particular, this research should be coupled with fracture mechanics studies of slow crack growth.

It also has to be noted that the experimental research has employed only constant or ramp load histories , the latter in rate–of–loading studies. Although the models have been applied to intermittent service loads, and could be used for cyclic–type loads, they remain to be experimentally verified for these more general cases.

Models are also required for creep and relaxation. Although the viscoelastic properties of wood are nonlinear and dependent on the moisture/temperature state, linear models resulting from combinations of springs and dashpots could be fitted to data as an approximation. However, the determination of whether the viscoelastic response is fluid (with a monotonic increase in strain under constant stress) or solid (upper bound on strain), presents the same difficulties as the threshold stress ratio for duration of load. In lieu of continuing tests for very long times, a model with solid behavior should be used to represent the data, obtaining from the regression an estimate of the delay in arriving at an upper bound for strain. A long delay would correspond to a low load duration threshold and vice versa.

6. Development of Other Adjustment Factors

The general adjustment factor K in Eq.(1) is not only composed of the duration of load adjustment, K_d, since it must also include a factor K_s for other service conditions of moisture or temperature. In addition, it could include adjustments for strength or stiffness changes due to chemical treatments. Since duration of load and moisture effects are linked, it is only a simplifying assumption that the overall adjustment can be expressed as the product $(K_d \cdot K_s)$, with K_s derived from considerations of only the short–term strength. Nevertheless, using this assumption, what is required is a behavior model for the effect of moisture content on short–term strength. Studies of this kind have produced data on strength changes when the moisture content has been varied from one equilibrium state to another. It has been observed that strength changes are more severe for stronger material, and less pronounced at the lower tail of the short–term strength distribution. Furthermore, strength generally increases with decreasing moisture content, but severe drying may produce an strength decrease.

In service, moisture changes may not reach equilibrium conditions, particularly for heavier specimens, and what would be needed for reliability studies is a comprehensive model linking non–steady state moisture movements with strength degradation, under actual environmental conditions over the service life. The latter are again samples from an stochastic process, as in the case of snow loads. There is a need for research in this area, both at the theoretical as well as the experimental level. There have been, however, experiments where moisture effects have been assessed under service conditions, with specimens exposed to the natural variations in the environment for a period of time (Madsen, 1980). It may also be argued that, since the duration of load data on lumber were obtained in semi–controlled, sheltered environmental conditions, the results already reflect the combined adjustment for duration of load and mild environmental variations. In the case of the hemlock lumber experiment, the temperature was controlled to be above 20 °C, but the relative humidity changed freely from a minimum of about 30% in the winter to a maximum of about 80% in the summer. Peaks, however, did not last more than 2 weeks.

For the moment, there can only be a simplistic answer to this complicated problem. If the characteristic short–term strength is given for a base equilibrium moisture content, it can be adjusted with K_s for other equilibrium conditions. The duration of load adjustment, K_d, derived for the base conditions, is assumed to apply equally to the modified situation. Strictly speaking, the adjustment K_s must be obtained by a procedure entirely similar to that described for K_d in Figure 7. Curve I would now represent the $\beta - \phi$ relationship when duration of load is ignored and the strength is assumed random, but at the base moisture, while the applied load is random and equal to the maximum over the service life. Curve II would give the $\beta - \phi$ relationship when duration of load is also ignored, but now the short–term strength is varying responding to stochastic changes in moisture during the service life. Obviously, we need a model to obtain Curve II, similar in a sense to the damage accumulation model for load duration, with simultaneous consideration of two stochastic processes: one for moisture content and another for applied loads. The factor K_s would result, as before, from the ratio

$$K_s = \phi_{II}/\phi_I \tag{8}$$

calculated at the target reliability level. Recently, Barrett et al. (1990) have followed this approach to determine K_s, but making the simplifying assumption that the short–term strength, when changed from the base to a different moisture content, remains constant during the service life under the application of the load. Although this is an unrealistic assumption, the study is useful in that if follows the proper way for the determination of K_s.

Similar procedures have to be implemented for adjustments due to temperature variations, also stochastic during the service life. On the other hand, chemical treatments are only applied once and adjustment factors would follow from consideration of the corresponding changes in short or long–term behavior.

7. Dynamic Characteristics

Reliability of wood structures under dynamic loads, particularly earthquakes, is only recently receiving attention. Contrary to a long–standing assumption, wood structures may not be necessarily inmune to great damage or collapse during an earthquake. While it is true that inertia forces in lightweight wood structures are less than in heavy masonry or concrete buildings, the nature of the non–rigid connections used with wood and their degradation during shaking needs to be carefully studied. As for other materials, characteristics such as damping and hysteretic behavior are required. In particular, it is necessary to study the hysteretic behavior of connectors (by experiment or by modelling), and to determine the pinching of the hysteresis loop resulting from connection degradation, loosening and reduced stiffness. This pinching is influenced both by the nonlinear behavior of the wood in compression during loading and unloading, and the propagation of cracks parallel to the fibers in a combined fracture mode.

Although stresses induced during an earthquake vary over time, it is reasonable to assume that collapse may follow from a case of overload rather than accumulation of damage produced by lesser loads over a long time. Thus, duration of load effects may be ignored for short–term, infrequent situations like earthquakes.

Another important problem within the dynamic response of wood structures is that of serviceability under vibration limit states, particularly for lightweight, long span floors. Again, the reliability formulation requires quantification of the demand and an understanding of the capacity (in this case human tolerance to vibration levels).

8. Conclusions

The application of reliability–based procedures to the design or assessment of wood structures hinges on a detailed knowledge of the material behavior. From the few topics discussed, it is apparent that this behavior reflects a complicated material and that our knowledge, although greatly advanced in recent years, still lacks in several areas. Although experimentation must be central to new progress, it must be coupled with more fundamental studies at the theoretical level, resulting in models which can be used in reliability evaluations under different service conditions.

9. References

Barrett, J.D. and Lau, W. (1990). "A Comparison of Deterministic and Reliability Based Moisture Content Adjustment Factors", Proceedings, 1990 International Timber Engineering Conference, Tokyo, Japan.

Foschi, R.O., Folz, B. and Yao, F. (1989). "Reliability-Based Design of Wood Structures", Structural Research Series Report No. 34, Department of Civil Engineering, University of British Columbia, Vancouver, B.C., Canada.

Foschi, R.O. and Barrett, J.D. (1982). "Load Duration Effects in Western Hemlock Lumber", Journal of the Structural Div., ASCE, Vol. 108, No.7.

Hoffmeyer, P. (1990). "Failure of Wood as Influenced by Moisture and Duration of Load". Ph.D. Dissertation, Environmental Science and Forestry, SUNY, Syracuse, New York.

Madsen, B. (1973). "In-Grade Testing: Problem Analysis", Forest Products Journal, Vol. 28, No. 4.

Madsen. B. (1980). "Moisture Effects in Lumber", Structural Research Series Report No. 27, Department of Civil Engineering, University of British Columbia, Vancouver, B.C., Canada.

Madsen. B., and Barrett, J.D. (1976). "Time-Strength Relationship for Lumber", Structural Research Series Report No. 13, Department of Civil Engineering, University of British Columbia, Vancouver, B.C., Canada.

Nielsen, L. (1980). "Stress-Strength-Lifetime Relationship for Wood", Wood Science, Vol. 12.

THE DEVELOPMENT OF LSD CODES FOR STRUCTURAL TIMBER

R. H. LEICESTER
Chief Research Scientist
CSIRO Division of Building, Construction and Engineering
P.O. Box 56
Highett, Victoria 3190
Australia

ABSTRACT. This paper contains a discussion on the various considerations and procedures that are involved in the development of LSD codes for timber engineering. It includes matters related to format, determination of characteristic values, code calibration and the application of structural reliability techniques. The discussion also includes some legal considerations and basic strategies for code development.

1. Introduction

1.1 CODE EVOLUTION

Limit states design, LSD codes (which typically are reliability based) are a natural evolution from working stress design, WSD, codes. Both codes present design rules in a deterministic format. The essential difference is that for LSD codes, both design resistances and loads are expressed in terms of partial safety factors. The system of partial safety factors provides a means for obtaining improved consistency in structural reliability.

In addition to the above, the formats of LSD codes usually contain a careful definition of the design limit states for which specific design rules are given. For example, the draft Eurocode 5 (European Committee for Standardisation 1990a) includes reference to the following limit states:

Ultimate limit states
– rupture of structural members
– loss of equilibrium of the structure considered as a rigid body

Serviceability limit states
– deformations that affect appearances
– discomforting vibrations.

In their technical content, LSD codes usually exhibit considerable improvements on their WSD predecessors. Partly this is due to the extensive advances in reliability-based procedures for the design of engineered timber structures and timber engineering technology during the past two decades and the pressure from technologists to include this information in drafting the design rules of the new LSD codes. The improvements, however, are also partly due to actions that are motivated by two legal considerations. The first is the pressure to use the newly developed reliability techniques for assessing the validity of design recommendations. The second is the fact that the characteristic

J. Bodig (ed.), Reliability-Based Design of Engineered Wood Structures, 91–124.

values used in LSD codes are parameters that are directly measurable; hence there is a new tendency to resort to full-size testing, either to define material properties or to resolve disputes related to structural quality. By contrast, it is to be noted that with WSD codes, the specified design parameters cannot be measured directly and hence in a dispute involving WSD codes, there is always scope for arguments concerning the magnitude of the factors that should be applied in processing test data.

The following is a discussion on the considerations and procedures that are involved in the development of LSD codes. Because much of the discussion is based on private communications from code committee members, and because much of it relates to draft codes, no attempt is made to provide a comprehensive catalogue and comparison of the LSD codes of various countries. Rather, the intention is to use information from the published and draft codes to provide illustrative examples of the alternative approaches that are available for drafting an LSD code.

Most of the discussion relates to the published timber engineering LSD codes of Canada (Canadian Standards Association 1989), Denmark (Danish Standards Institute 1983), and to draft codes of Australia (Leicester 1990), Eurocode 5 (European Committee for Standardisation 1990a), New Zealand (Standards Association of New Zealand 1991) and the USA (Goodman 1990). Each of these codes is associated with a set of auxiliary standards which are used for classifying structural components and other design codes such as loading codes. Background information on drafting these codes can be obtained from Foschi *et al.* (1989), Larsen (1984), Walford (1989), Leicester *et al.* (1986), Leicester (1990), Ellingwood *et al.* (1980), Goodman (1990) and Gromala *et al.* (1990).

1.2 THE BASIC FORMAT

For structural timber elements, the design member capacity R* in LSD codes is usually stated in the form

$$R^* = k_{cum} R_k/\gamma_m \tag{1}$$

or

$$R^* = \phi\, k_{cum} R_k \tag{2}$$

where:

- R_k = characteristic value (typically a five-percentile value)
- k_{cum} = a cumulative modification factor for various design conditions
- γ_m = 'partial safety factor' for the structural resistance
- ϕ = 'resistance factor' or a 'capacity reduction factor'.

The factors γ_m and ϕ are equivalent parameters with a relationship roughly given by $\phi = 1/\gamma_m$ for normal design situations. The safety factor, γ_m, tends to be used in European countries, while the resistance factor, ϕ, is employed elsewhere.

The cumulative modification factor is usually given by

$$k_{cum} = k_1.k_2.k_3\ldots \tag{3}$$

where $k_1,k_2,k_3\ldots$ refer to modification factors such as k_D for load duration effects, k_{SH} for load sharing or system effects, and k_{INST} for instability or buckling effects.

Equations (1) and (2) represent the basic format used for design resistance. A similar format is used for design stiffness. In some cases, a design resistance may be expressed as a function of both factored characteristic strengths and/or stiffnesses.

The design load effect Q^* is generally a linear sum of load effects Q_i arising from various types of loads, i.e.

$$Q^* = \Sigma \gamma_i Q_i \tag{4}$$

where:

γ_i = partial safety factor.

As an illustrative example of Eq. (4), examples are taken from the Australian Standard AS 1170.1 (Standards Australia 1989a) for load combinations involving dead load effect, D, floor live load effect, L, and wind load effect, W.

For ultimate limit states in which the dead load acts to increase the total load effect,

$$Q^* = 1.25\,D + 1.5\,L \tag{5a}$$

$$Q^* = 1.25\,D + 0.4\,L + W_u \tag{5b}$$

where the dead load effect, D, denotes an estimate of the average value, and the live load effect, L, and wind load effect, W_u, denote loads that have a 5 per cent chance of exceedence for a randomly chosen building within any 50-year period; thus the wind load, W_u, is based on an estimate of the 1000-year return wind gust.

For ultimate limit states in which the dead load acts to reduce the total load effect,

$$Q^* = 0.8\,D + 1.5\,L \tag{6a}$$

$$Q^* = 0.8\,D + 0.4\,L + W_u \tag{6b}$$

Note that a lower bound of the dead load effect, i.e. 0.8 D, is now used. For the case of fire limit states

$$Q^* = 1.1\,D + 0.4\,L. \tag{7}$$

For all these cases, the design equation is given by

$$R^* \geq Q^*. \tag{8}$$

For the case of rigid body motion, AS 1170.1 also makes use of Eq. (6), so for this case

$$0.8\,D + 1.5\,L \geq 0 \tag{9a}$$

$$0.8\,D + 0.4\,L + W_u \geq 0. \tag{9b}$$

For designing to meet specified serviceability limit states, the load combinations given for ultimate limit states may be used when the target risk of failure is small. When the acceptable risk of failure is relatively large or frequent, then use may be made of two special sets of load combinations.

For short-term effects (such as building sway), the specified serviceability load combination is one which has a 5 per cent chance of exceedence in any one year; some typical examples are

$$Q^* = D + 0.7\,L \tag{10a}$$

$$Q^* = D + 0.4\,L + W_s \tag{10b}$$

where:

W_s = wind load effect based on a 20-year return wind gust.

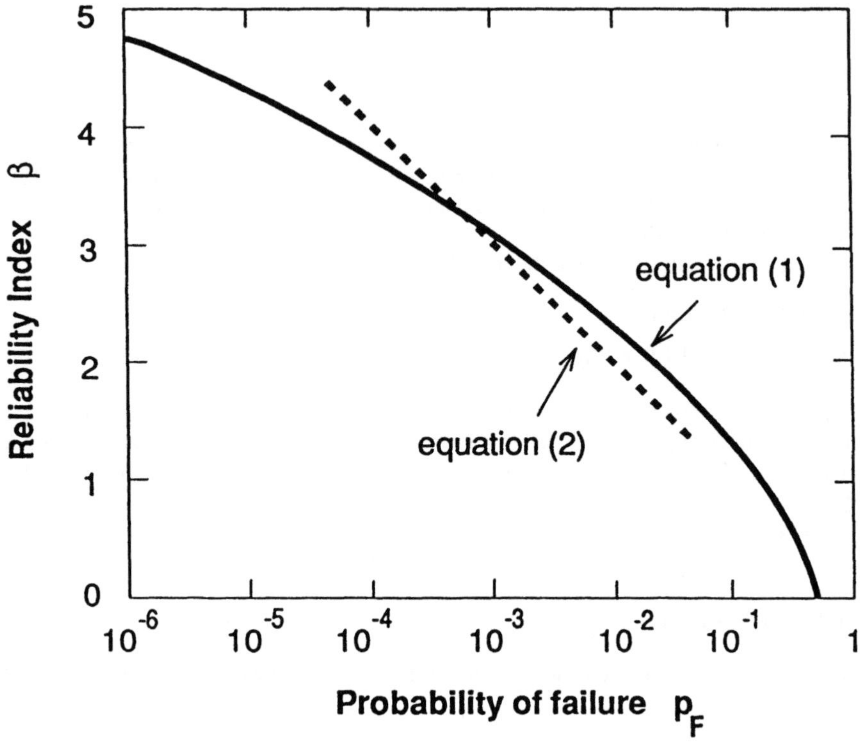

Figure 1. **Graph of the Reliability Index.**

For long-term serviceability effects (such as those related to creep), the design load combinations refer to loads averaged over the lifetime of a building; specifically the design load combination is chosen so that the chance of exceedence in a randomly chosen building is 5%. A typical example of this load combination is:

$$Q^* = D + 0.4\,L. \tag{11}$$

The basis of Eqs (10) and (11) has been described by Pham and Dayeh (1986).

All of the above demonstrate the great variety of design load effects that are considered in LSD codes.

2. Structural Reliability

2.1 USE BY CODE COMMITTEES

Many members of LSD code committees have had little or no experience in the use of structural reliability techniques. As a result, they found it difficult to make use of some of the more sophisticated reliability concepts for committee decision purposes. However, there is a surprising amount of reliability theory that can be used by a committee to assist in resolving difficult decisions. Some of this is discussed next.

2.2 COMPLEX RELIABILITY CONCEPTS

There would appear to be a general, if reluctant, acceptance of the concept of a reliability index, β, defined by:

$$\Phi(-\beta) = p_F \tag{12}$$

where:

p_F = the probability of failure in-service

Φ = the cumulative distribution function of a standardised normal variate.

The definition of β, as given by Eq. (12), appears to be academic to a structural engineer without formal training in reliability techniques. However, an engineer is usually prepared to accept it, because it is related in some rough way to the estimated probability of failure, a valid yardstick of the success of his design projects. However, there is little doubt that the concept of a reliability index would be far easier to accept if the definition of β was replaced by a simple empirical polynomial or by some other simple function such as

$$\beta = -\log_{10}(p_F). \tag{13}$$

Equations (12) and (13) are graphed in Fig. 1 and tabulated in Table 1.

For committee discussion purposes, the use of complex failure functions appear to be acceptable provided all parameters are represented by simple static random variables, i.e.:

$$p_F = \Pr\{f(X_1,X_2,X_3 \ldots X_n) < 0\} \tag{14}$$

where:

$f(X_1,X_2 \ldots X_n)$ = general, possibly non-linear, function of the static random variables $X_1,X_2 \ldots X_n$.

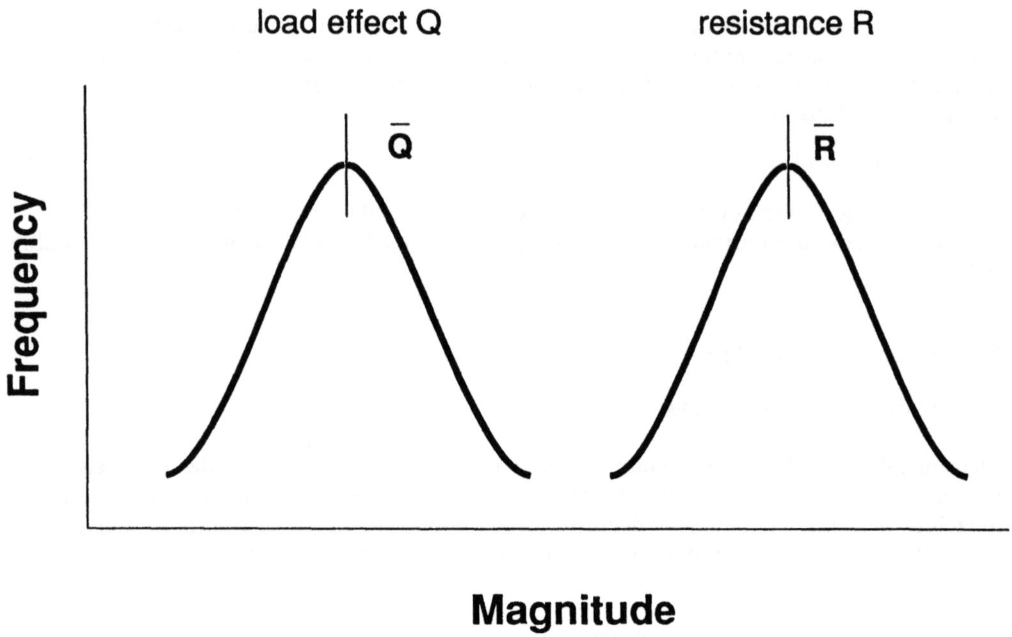

Figure 2. Statistical Models for Simple Distributions.

Ideally the random variables used should be well-known two-parameter distributions, e.g. Weibull, Gumbel, gamma and lognormal. Fortunately, a simple algorithm is available for obtaining a good approximation to the solution of Eq. (12) (e.g. Leicester 1985a, Foschi *et al.* 1989).

The implications of the parameters involved in a time-varying process are not easily understood by a non-specialist committee. However, through the use of 'Turkstra's Law', such processes may be represented approximately by pairs of static random variables (Ellingwood *et al.* 1980, Leicester *et al.* 1986a). For example, if the time-varying variable Y(t) is given by

$$Y(t) = X_1(t) + X_2(t) + X_3(t) \tag{15}$$

where $X_1(t)$, $X_2(t)$ and $X_3(t)$ are time-dependent variables, then the peak value of Y(t) within a specified reference period is given by the greatest of the following,

$$Y_{peak} = X_{1,peak} + X_{2,apt} + X_{3,apt} \tag{16a}$$

$$Y_{peak} = X_{1,apt} + X_{2,peak} + X_{3,apt} \tag{16b}$$

$$Y_{peak} = X_{1,apt} + X_{2,apt} + X_{3,peak} \tag{16c}$$

where the subscripts 'peak' and 'apt' denote the peak and arbitrary point-in-time values. Note that all X and Y in Eq. (14) refer to static random variables. Thus, each time-dependent variable X(t) is replaced by a pair of static random variables X_{peak} and X_{apt}.

An interesting application of this law may be found in the derivation of loads for fire limit states. Here, the fire is taken to be the peak event, while the design load effect such as that given by Eq. (7) is roughly an arbitrary point-in-time value.

2.3 SIMPLE RELIABILITY CONCEPTS

Because of their simplicity, approximate solutions that are stated solely in terms of mean values and coefficients of variation, are very useful for committee discussion purposes, and for interpolation purposes in drafting code design rules.

As an example, when consideration of the structural resistance and load effect can be reduced to the equivalence of two lognormal random variables R and Q as illustrated in Fig. 2, then the reliability index, β, is given approximately by:

$$\beta \approx \ln(R_{mean}/Q_{mean})/(V_R^2 + V_Q^2)^{1/2} \tag{17}$$

where:

R_{mean} = mean value of resistance
Q_{mean} = mean value of load effect
V_R = coefficient of variation of resistance
V_Q = coefficient of variation of load effect.

Equation (17) is too crude to be used directly for codification purposes; however it does serve to illustrate the effect on the factor of safety R_{mean}/Q_{mean} in response to changes in the design uncertainty $V_R^2 + V_Q^2$. The following is an application that is more useful.

Ignoring the cumulative modification factor, k_{cum}, the design resistance, R*, for the simple problem given in Fig. 2 may be defined by

$$R^* = \phi R_{0.05} \tag{18}$$

where:

$R_{0.05}$ = five-percentile value of strength.

Equations (17) and (18) lead to the approximation (Ravindra and Galambos 1978)

$$\phi = k_{com} (R_{mean}/R_{0.05}) \exp(-0.6\, \beta\, V_R) \tag{19}$$

where:

k_{com} = a 'committee factor', an arbitrary factor by which both specified loads and resistances are multiplied.

It is assessed that in Australian limit states codes, a value of $k_{com} = 0.9$ is being applied. Equation (19) is useful for interpolation purposes.

Most practical design situations can be examined with such simple approximations. To do this, the statistics of complex design functions are simplified through the application of Taylor's series expansions. For example, a load effect, Q, that is a non-linear function of a load represented by a random variable Z, i.e.

$$Q = f(Z) \tag{20}$$

has a mean value given approximately by

$$Q_{mean} \approx f(Z_{mean}) \tag{21}$$

and a coefficient of variation given approximately by

$$V_Q \approx [f'(Z_{mean})/f(Z_{mean})]V_Z \tag{22}$$

where:

Z_{mean} = mean of Z
V_Z = coefficients of variation of Z
$f'(\,)$ = first differential of $f(\,)$.

Equation (22) can be used to examine the effects of non-linearity on variability.

2.4 TARGET RELIABILITIES

Target reliabilities may be selected on the basis of social acceptance. For example, Reid (1989) undertook a survey on the literature of risk and concluded that the socially acceptable risk of untimely death through structural failure is about 10^{-5} per person in a 50-year period.

On the basis that the average person spends somewhat less than one-tenth of his life being supported by critical structural elements, but that there are also other socially undesirable consequences of a structural failure apart from loss of life, the target reliability for critical structural elements is about $p_F = 10^{-4}$ or $\beta = 4$ for a 50-year reference period.

Another useful basis for selecting target reliabilities is to make use of cost-optimisation concepts (e.g. Leicester and Beresford 1977, Leicester 1984). For the case of ultimate limit states, the optimised probability of failure, $p_{F,opt}$, is given roughly by

$$p_{F,opt} \approx 0.5\, V_R(C_R/C_F) \tag{23}$$

where:

C_R = cost of the structural element
C_F = effective costs that would be incurred should an in-service failure occur.

Such effective costs would include values for loss of life, injury, loss of business and damage to a professional reputation. As an example, values of $V_R = 0.3$ and $C_R/C_F = 0.001$ lead to $p_{F,opt} = 0.00015$ or a reliability index of about $\beta = 4$.

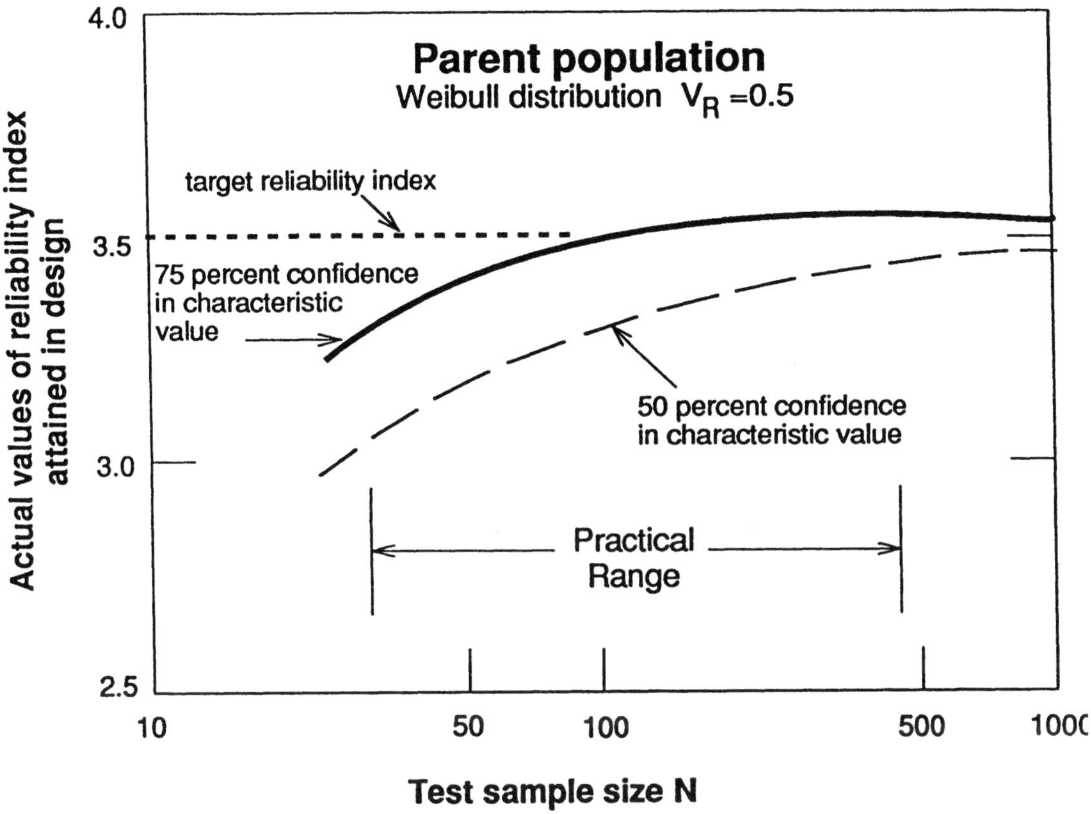

Figure 3. **Effect of Choice of Confidence Level used in Sampling for a Five-percentile Characteristic Value (after Leicester 1986a) when the Target Reliability Index is 3.5.**

Equation (23) reveals the interesting feature, often observed in reliability studies, that because the cost of failure is relatively higher, a smaller risk is more appropriate for connections when compared with solid timber members in the same structure. It should be emphasised that these target reliabilities refer to in-service failure rates. As will be noted later, other values of target reliabilities may be more appropriate in risk calibrations because of difficulties associated with making accurate risk assessments.

2.5 SAMPLING

In the derivation of partial factors of safety, due account should be taken of the uncertainties associated with obtaining information from limited sample sizes. Fortunately some procedures are available that are simple enough for use in code applications.

If a characteristic value is estimated from a limited sample size, N, then the estimate will, in general, be either higher or lower than the true value of the parent population. If the estimate is lower than the true value for 75 per cent of all samples, then the estimation algorithm is said to give a 75 per cent confidence against an overestimate. As an example, if the targeted characteristic value of strength, R_k, is the five-percentile value, $R_{0.05}$, then the following algorithms will give the estimate with a 75 per cent confidence against an overestimate (Leicester 1986a, 1986b, 1987)

$$R_k = R_{0.05,npe}\,[1 - 2.7\,V/\sqrt{N}] \quad \text{for } N \geq 20, \text{ and} \tag{24}$$

$$R_k = R_{min}\,(N/27)^V \quad \text{for } N < 20 \tag{25}$$

where:

- N = sample size,
- $R_{0.05,npe}$ = non-parametric estimate of the five-percentile value read off directly from the ranked data set,
- R_{min} = minimum value in the sample,
- V = coefficient of variation of the data.

The value of V may be estimated from prior experience in the case of small sample sizes.

In the case of an estimate with 50 per cent confidence:

$$R_k = R_{0.05,npe} \quad \text{for } N \geq 15, \text{ and} \tag{26}$$

$$R_k = R_{min}\,(N/13.5)^V \quad \text{for } N < 15 \tag{27}$$

Equations (24)–(27) illustrate the effects of various parameters on estimates from limited samples. It is apparent that the choice of optimum sample size will be based in part on the coefficient of variation, V.

The effect of choice of confidence level on the relationship between the targeted and attained safety levels has been studied (Leicester 1986a) and significant differences were found when small sample sizes and large coefficients of variation occur simultaneously. An example of this is illustrated in Fig. 3.

2.6 APPLICATION

In the application of reliability theory as part of formal code committee operations, it is important that the committee agree on several matters. First it is necessary to choose a reference period to which the probability of failure is to be related. For a fixed set of design rules, the computed failure rate will increase as the design period increases. However, extrapolation of failure rates from one

TABLE 1. Values of the Reliability Index

p_F	Reliability index, β	
	Equation (11) $\beta = -\log_{10}(p_F)$	Equation (12) $\Phi(-\beta) = p_F$
10^{-2}	2.0	2.33
10^{-3}	3.0	3.09
10^{-4}	4.0	3.72
10^{-5}	5.0	4.26

TABLE 2. Statistics of the Structural Analysis Parameter (After Leicester *et al.* 1986)

Load type	H_{mean}/H_{nom}	cov(H)
Dead load	0.90	0.10
Floor live load	0.85	0.15
Wind load	0.80	0.20
Wind load if designer ignores the reinforcing effects of cladding	0.60	0.25

H = structural analysis parameter to convert a load, S, to a load effect, Q, where Q = HS.
H_{mean} = mean value of H.
H_{nom} = value of H corresponding to the nominal value given by a typical engineering analysis.

TABLE 3. Components of Variability due to the Uncertainties of Structural Theory of Strength (After Pham and Leicester 1979)

Origin of variability	Typical coefficients of variation
For one particular theory of analysis	0.1 – 0.3
Between various theories of analysis	0.1 – 0.3
Due to application of one theory to various types of structures	0.15 – 0.35
Due to differing interpretations of one theory by various designers	0.2 – 0.5
Due to the use of discrete sizes for structural elements	0.05 – 0.15

reference period to another is difficult because some loads (such as dead loads) are correlated in value from one year to the next, whereas other loads (such as wind loads) are almost uncorrelated. Typical examples of design periods currently used for ultimate limit states are 50 years for Australia (Leicester *et al.* 1986) and 30 years for Canada (Foschi *et al.* 1989). In some countries a reference period of one year has been proposed.

The second important matter to decide is the definition of failure. With respect to ultimate limit states, there appears to be general agreement that this refers to single elements rather than total structural systems. An exception is that the collapse of total structural systems is often considered to be the failure criterion when analyses are undertaken to evaluate system effects (Foschi *et al.* 1989).

Another matter to be considered is whether computed failure rates are intended to be predictions of real values, or whether they are intended to be purely nominal values based on idealised load and strength models. Both approaches have been used in code assessments. The advantage of attempting to compute realistic failure rates is that this realism encourages acceptance by code committees and also that it is more effective in exposing parameters that have a significant influence on structural reliability.

In attempting to compute realistic failure rates, due account must be taken of the structural influences of 'non-structural' components such as internal partitions, and facade and roof claddings. In addition, due account must be taken of the uncertainties associated with the structural theory of strength, and the uncertainties associated with the structural theory that converts loads (such as wind pressures) to load effects (such as column moments). Examples of the statistical parameters used by Australian code committees for these purposes are given in Tables 2 and 3.

3. Code Calibration

3.1 SOFT CONVERSION

The simplest method for calibrating LSD codes is by soft conversion of an existing WSD code. The procedure is based on the aim of obtaining an LSD code that produces designs that are on average identical with those of the WSD code.

For example, in Australia the specified design load combination for an LSD code has a magnitude on average of about 1.35 time that for a WSD code. Hence, in a soft conversion, a material coefficient, ϕ, is derived from

$$\phi\, k_{cum,LSD}\, R_k = 1.35\, k_{cum,WSD}\, R_{nom} \tag{28}$$

where:

$k_{cum,LSD}$ = cumulative modification factor for LSD codes
$k_{cum,WSD}$ = cumulative modification factor for WSD codes
R_{nom} = resistance, based on basic working stresses in a WSD code.

Usually in a soft conversion all the modification factors, $k_1, k_2 \ldots$, remain the same except that the duration factor, k_D, is now taken to be 1.0 in the case of loadings of 3 to 5 minutes duration. For the Australian codes this is

$$k_{D,WSD} = 1.75\, k_{D,LSD} \tag{29}$$

where:

$k_{D,WSD}$ = duation factor for WSD codes
$k_{D,LSD}$ = duration factor for LSD codes.

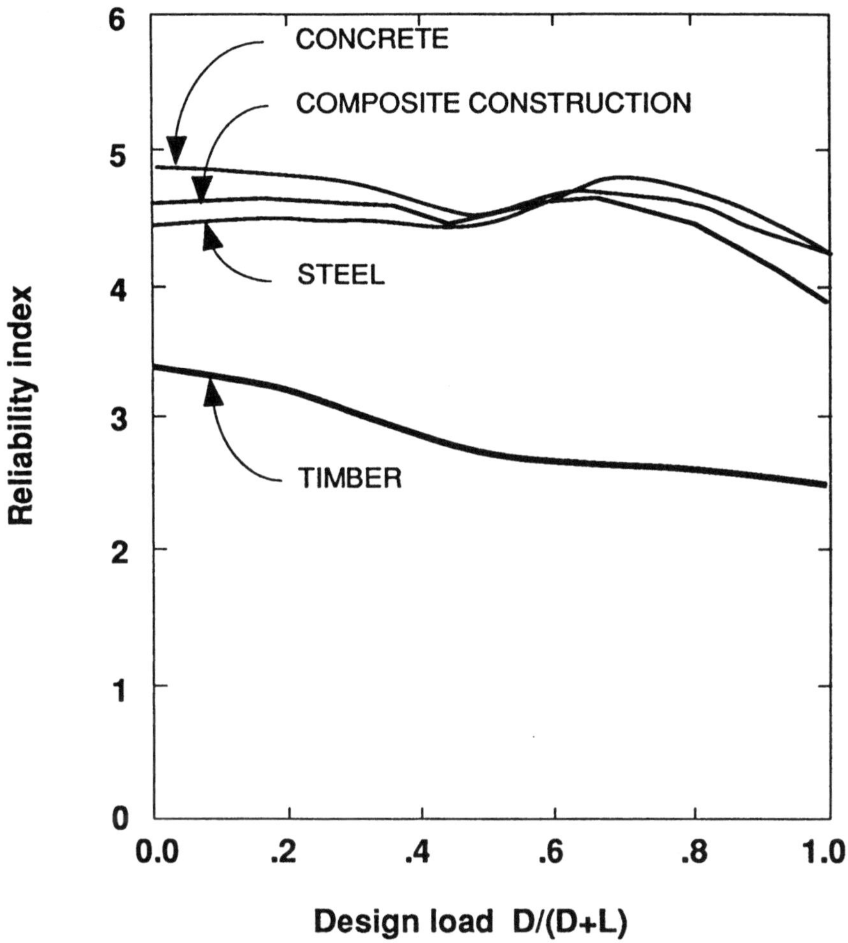

Figure 4. **Reliability Indices for Beams Designed According to Australian LSD Codes (after Leicester et al. 1986).**

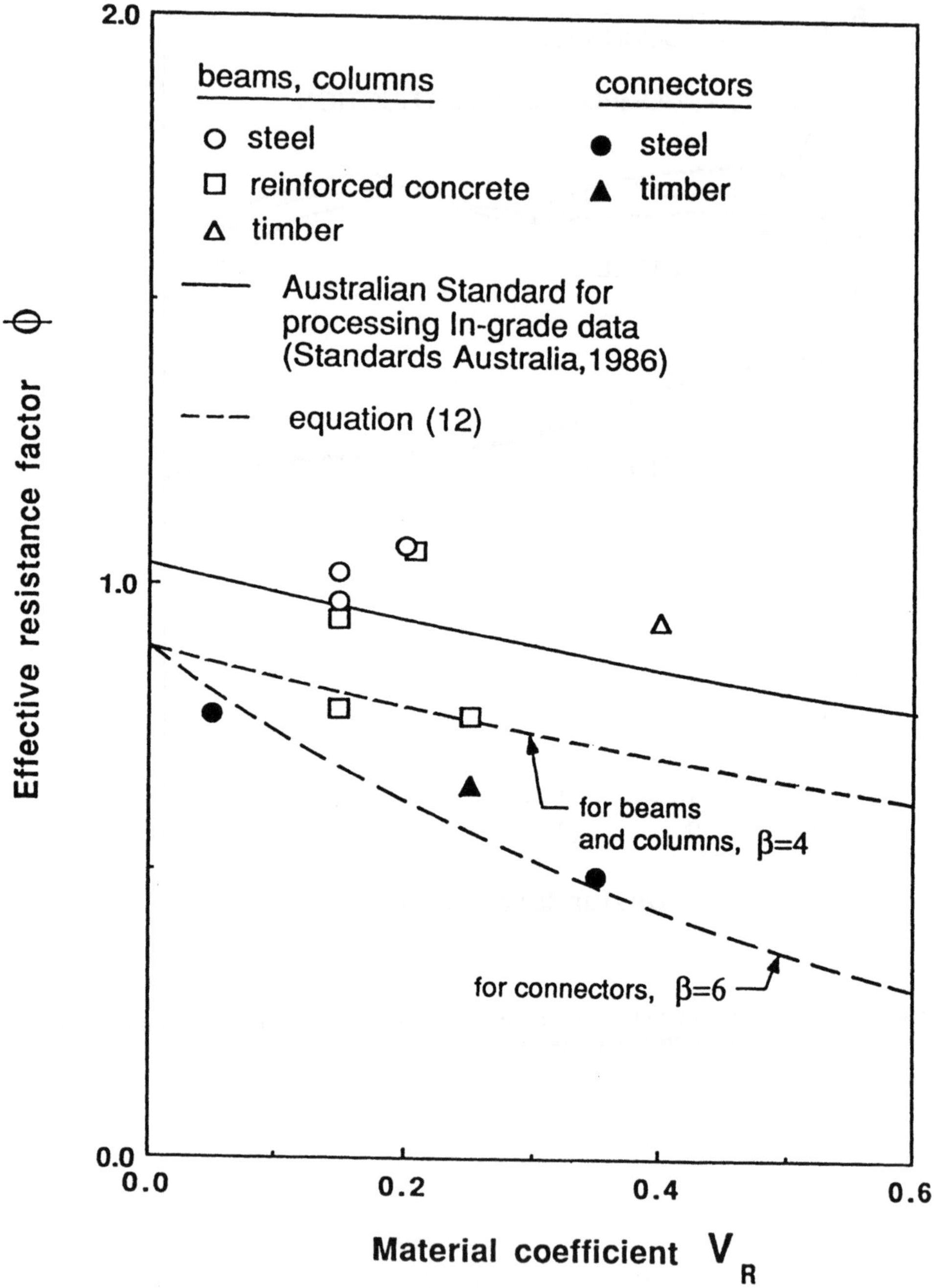

Figure 5. **Resistance Factors on the Five-percentile Capacity for Australian LSD Codes (after Leicester 1990).**

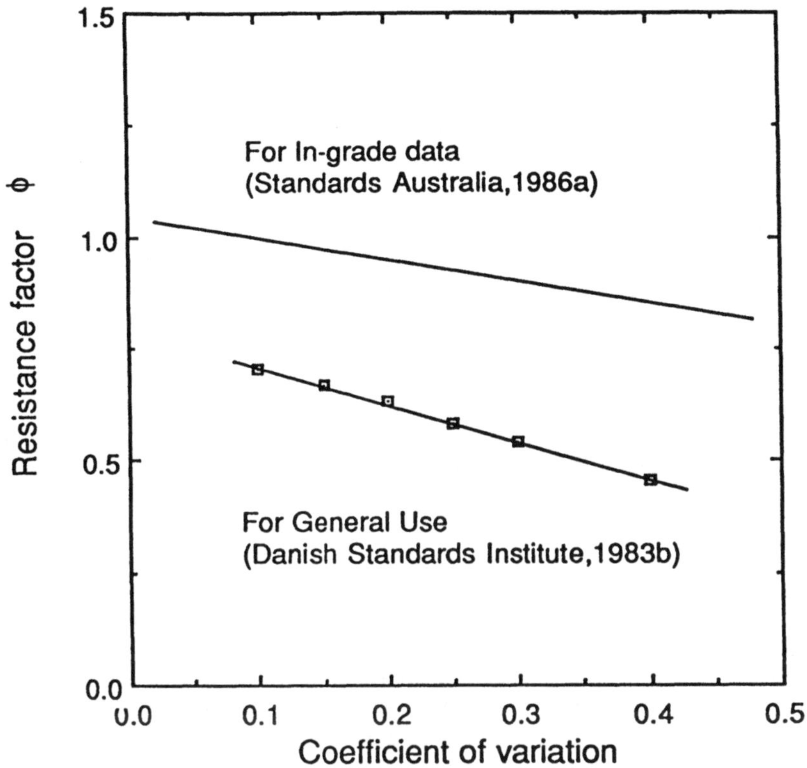

Figure 6. Resistance Factors Specified in Standards.

Equations (26) and (27) then lead to

$$\phi \, R_k = 2.36 \, R_{nom} \qquad (30)$$

A similar equation has been derived by Foschi *et al.* (1989) for the soft conversion to the Canadian LSD code CAN3–086.1–M84 (Canadian Standards Association 1984).

3.2 HARD CONVERSION

In a hard conversion to an LSD code, little consideration is given to prior recommendations of existing WSD codes. Rather, the design recommendations are derived directly from reliability studies, with probabilities of failure targeted to absolute values, such as those indicated earlier in Section 2.4 of this paper.

The calibration of an LSD code through a pure hard conversion operation is usually not feasible in the case of ultimate limit states because the available data are inadequate to compute the probability of failure with sufficient accuracy. However, in the case of serviceability limit states where higher failure rates are the optimum, it should be possible to use a hard conversion. Simple reliability models for this purpose have been discussed previously (e.g. Leicester and Beresford 1977, Leicester and Pham 1987).

3.3 RELIABILITY ASSISTED CALIBRATIONS

Although assessment of the exact probability of failure is not feasible in the case of ultimate limit states, it is still possible to use reliability models in a comparative sense. Thus, a very effective procedure for calibrating LSD codes is to use reliability concepts for interpolation purposes or for obtaining consistency between various design situations. To do this, a basic set of assumptions is selected by the code committees for the reliability analyses and then when complete the results of these analyses are displayed to the committee as a basis for decisions. Two examples of such displays are shown in Figs 4 and 5.

In Fig. 4, reliability indices computed according to Eqs (12) and (14) for Australian LSD codes are used to assess the effects of various gravity load combinations and to compare LSD codes for timber with those of other materials. In Fig. 5, the first order approximation given in Eq. (19) is used to provide a comparative assessment of the reliability associated with the choice of resistance factors for Australian LSD codes.

4. Characteristic Values

4.1 CHOICE OF PERCENTILE

There would appear to be general acceptance of the choice of the five-percentile value for characteristic strength and the mean value for characteristic stiffness. In the Australian draft standard (Standards Australia 1986a), a characteristic value of 1.4 $E_{0.05}$ is used as an alternative to E_{mean} for species or species mixtures that have a large coefficient of variation.

There is also a general acceptance of the fact that the resistance factor should reduce with increasing coefficient of variation, as illustrated by the examples in Fig. 6; this would appear to suggest that a value lower than the five-percentile value, perhaps the one-percentile value, is being targeted for the design strength. Thus, with hindsight, a suggestion could be made that the one-percentile value may have been preferable as the choice for characteristic strength; the uncertainties associated with an estimation of such a characteristic value would expose the true uncertainties of our knowledge of strength.

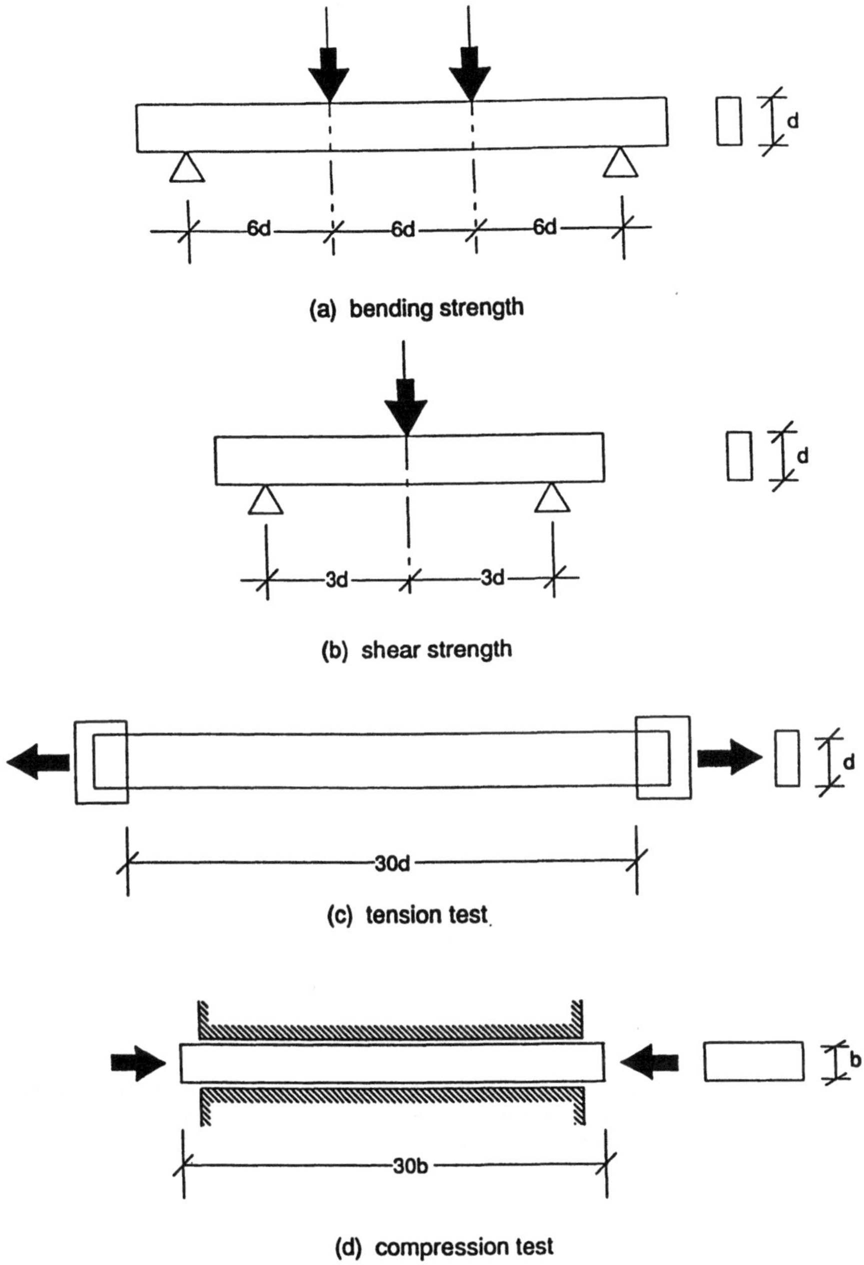

Figure 7. **Loading Configurations used in Australia for Evaluating the Characteristic Strength of Timber (Standards Association of Australia 1986a).**

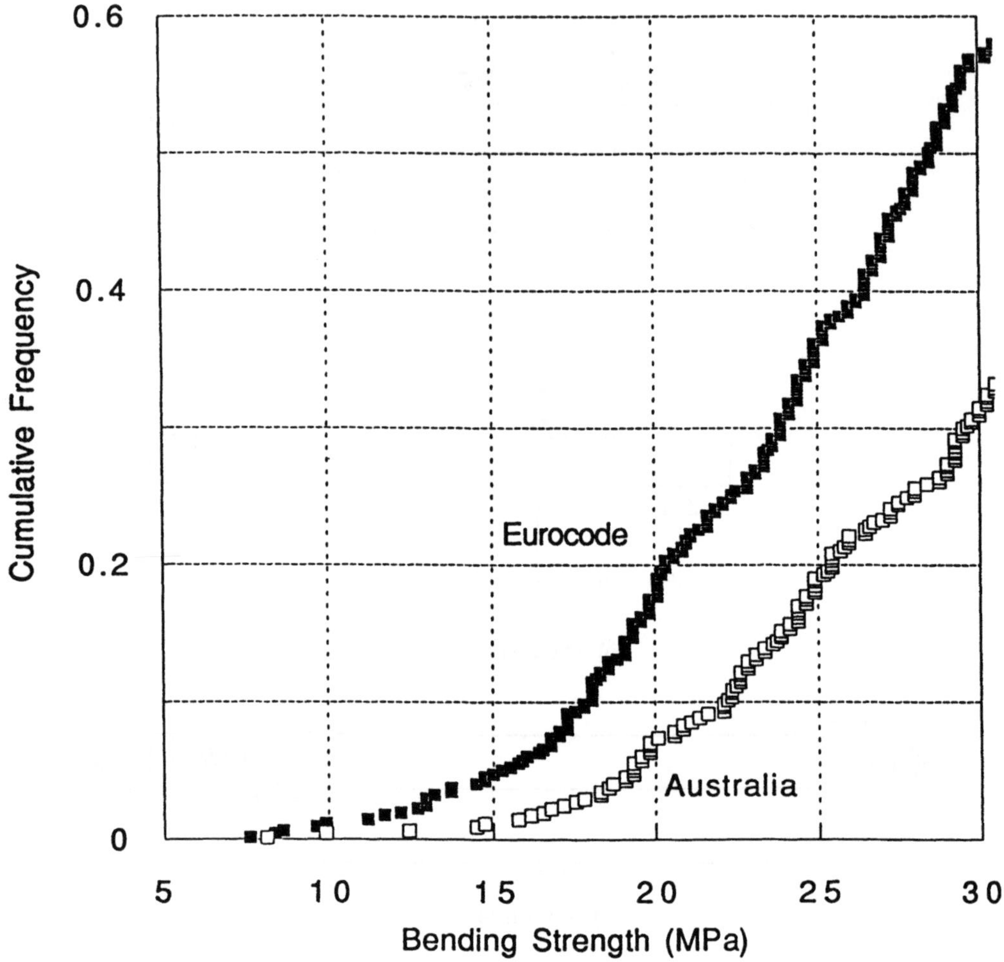

Figure 8. Bending Strength of F5-grade Radiata Pine as Measured by Australian and Eurocode Draft Standards (Standards Association of Australia 1986a, European Committee for Standardisation 1989).

4.2 SAMPLING

There is general agreement that samples used to measure characteristic values should be selected so as to be representative of the in-service populations of structural elements. In the strictest terms, this sampling should be wide enough to include not only material from all sources, but also material produced at various times. The requirements of sampling to measure fundamental characteristic values should not be confused with the special requirements of sampling to measure system effects; in the latter it is desirable to include special sequential sampling so that the effects of correlations of the properties of the timber within a single system can be determined.

4.3 TESTING

Measured values of structural properties can vary considerably with the choice of test procedure. For some countries, such as the USA, Canada and Australia, there is a philosophy that test specimens should be selected and tested more or less under the same conditions as they would be subjected to under service conditions. For example, in the case of structural timber, the draft Australian standard (Standards Australia 1986a) requires that test specimens be selected from random locations within a stick of lumber, and that the test configurations be as shown in Fig. 7.

Some standards, such as the European draft standard EN TC 124.202 (European Committee for Standardisation 1989) and the British standard BS 5280 (British Standards Institute 1979), do not attempt to simulate in-service conditions. For example, in the measurement of flexural properties, these standards require that a grade determining defect be placed near the centre of a beam, i.e. at the location of maximum stress. This procedure produces a negative bias which will vary with grade and species; such a bias should be taken into consideration either in the partial safety factor, γ_m, or in the cumulative modification factor, k_{cum}.

An example of the effect of test method on the derived bending strength is shown in Fig. 8. In this figure there is a comparison between the strengths as measured by Australian standards (Standards Association of Australia 1986b) and Eurocode standards (European Committee for Standardisation 1989). The two methods of measurement produce significantly different values of bending strength.

This difference creates a barrier to trade because there is no obvious equivalence between the characteristic strength values as measured by the two standards. Furthermore, there are difficulties in the transfer of technology, because the modification factors, k_1, k_2, k_3..., are factors related to specific definitions of bending strength. Similar comments relate to methods specified in standards for the measurement of stiffness.

The selection of test configuration is particularly important in the assessment of connector properties. Connectors can be used in a great variety of configurations and loaded in a variety of ways, each one being associated with many failure modes. For example, a bolted joint may be loaded in tension, bending, shear or torsion; it may be used to connect timber joined at various angles; the bolts themselves may be placed in numerous configurations; connector failures may be due to bolt bending, bolt tension, washer bearing, wood tension (parallel or perpendicular to grain), and wood compression (parallel or perpendicular to grain).

One approach would be to test every configuration for which design loads are to be specified; this is in line with the philosophy of the older type of standards such as ISO 8969 (International Standards Organisation 1990a). An alternative approach is to have some general theory for connector behaviour, such as the European yield theory for nailed and bolted joints, and then to develop standards for measuring the parameters to be used in the theory, such as the ISO draft N 140 (International Standards Organisation 1990b); an example of this approach is given in the discussion by Goodman (1990).

TABLE 4. Effect of Sample Size on Sampling Factor (After Eq. (22))

Coefficient of variation	Sampling factor for 75% confidence level*		
	N = 30	N = 100	N = 300
0.1	0.95	0.97	0.98
0.3	0.85	0.92	0.95
0.5	0.75	0.87	0.92

* Sampling factor = $1 - 2.7\,(V\sqrt{N})$.

TABLE 5. Specified Resistance Factors for North American LSD Codes (After Canadian Standards Association 1989, Goodman 1990)

Property		ϕ	
		Canada	USA
Sawn timber	– flexure	0.9	0.85
	– compression	0.8	0.90
	– tension	0.9	0.80
	– shear	0.9	0.75
Plywood	– all stresses	0.95	–
Connectors	– lateral load on nails	0.7	0.65
	– lateral load on bolts	0.6	0.65
	– lateral load on glulam rivets	0.6	0.65

TABLE 6. Specified Resistance Factors for New Zealand and Australian LSD Codes (aAter Standards Association of New Zealand 1991, Leicester 1990)

Structural element	ϕ
Sawn timber, poles and glulam – all stresses	0.8
Plywood and laminated veneer – all stresses	0.9
Connectors – nails	0.8
– other	0.7

4.4 CONFIDENCE LEVEL

In estimating a characteristic value it is necessary to use some confidence level against overestimation. The two most commonly used levels are the 75 per cent and the 50 per cent confidence levels; currently the 75 per cent level appears to be the favoured value.

The reliability implications of a choice in confidence level were discussed earlier in Section 2.5. The efficiency penalty associated with the choice of a 75 per cent confidence level is illustrated in Table 4. It indicates that the suitable sample size may range from 30 to 300, as the coefficient of variation of the element increases from 0.1 to 0.5.

5. Format

5.1 CHOICE OF RESISTANCE FACTORS

There would appear to be general agreement that in designing for serviceability limit states, the partial safety factor, γ_m, and its complement, the resistance factor, ϕ, are taken to be 1.0. However, different approaches are used for the selection of factors in the design of ultimate limit states.

In North America, Australia and New Zealand, the philosophy used is that the value chosen for the resistance factor, ϕ, should reflect the uncertainty of the specified strength. As indicated by Eq. (19), this would imply that ϕ should be a function of the coefficient of variation of strength; hence in the case of solid timber, ϕ would vary with species, grade, method of grading, size, property, etc. (Walford 1989). Carried to extremes, the numerous ϕ values required would mean that this concept would be unworkable, and hence the solution has been to use a limited number of normalised resistance factors. Some examples of normalised resistance factors are given in Tables 5 and 6. To compensate for this simplification, the true characteristic value, R_K, is replaced by a normalised value, $R_{K,norm}$, defined by

$$\phi R_K = \phi_{norm} R_{K,norm} \tag{31}$$

where:

ϕ_{norm} = normalised value of ϕ, i.e. the value of ϕ chosen for specification in an LSD code.

In Eurocode 5, there is a single partial safety factor for materials, $\gamma_m = 1.3$, which is reduced to the value of 1.0 in the case of design for accidental actions.

The generic Danish code DS 409 provides numerous factors to be considered in the selection of γ_m. These include:

- coefficient of variation of the material,
- consequences of failure,
- type of failure,
- accuracy of the design model,
- confidence level of material assessment, and
- quality control.

In the case of timber structures, these requirements have been interpreted in DS 413 as indicated in Table 7.

5.2 CAPACITY VERSUS STRESS FORMULATIONS

The following are some examples of capacity formulations for design strength. For a solid timber beam the design bending moment capacity M_u may be specified by

$$M_u = \phi_b k_{cum} F_{k,b} Z \tag{32}$$

where:

$F_{k,b}$ = characteristic bending strength of timber
Z = section modulus.

For composite constructions such as plywood, the choice of format is more complex. One type of specification for bending moment capacity is

$$M_u = \phi_b k_{cum} g_{ass} F_{k,bv} Z \tag{33}$$

where:

$F_{k,bv}$ = characteristic strength for the individual veneers
g_{ass} = assembly factor to account for the veneer lay-up used.

An alternative specification used in CAN/CSA–086.1–M89 is

$$M_u = \phi_b k_{cum} M_k \tag{34}$$

where:

M_k = specified characteristic moment capacity of a particular plywood lay-up.

Of the two formulations, that of Eq. (33) is most flexible, but that of Eq. (34) is easier to use for both design and for quality control purposes. There is of course no reason to prevent both formulations being used in the same code.

While a stress formulation is suited to the philosophy of WSD codes, the capacity (resistance) formulation is probably more appropriate for LSD codes. The capacity formulation is normally used by steel and reinforced concrete LSD codes, and it is useful in unifying the format for structural timber and connectors. There are however, some situations for LSD codes when a stress formulation may be preferable. These include check rules for the fracture strength of butt joints dispersed in a glulam beam, and rules for checking the capacity of elements in composite construction such as plywood-web box beams.

5.3 DURATION OF LOAD

It is generally agreed that the modification factor for the effect of duration of load on member strength, k_D, should be normalised so the $k_D = 1.00$ for 3–5 minute load durations. In this way the characteristic values of strength may be directly measured in laboratory tests.

For other load durations, two types of formats appear to be used. In some codes, such as AS 1720.1 (Standards Australia 1988), values of k_D are given for 'effective durations of peak loads', while for other codes, such as CAN/CSA–086 (Canadian Standards Association 1989), the values of k_D are applied to qualitative descriptions of load actions. Neither method is completely satisfactory. Fortunately, all codes also include specific guidelines as to the appropriate value of k_D to be used for loads specified in existing loading codes.

5.4 SYSTEM EFFECTS

The modification factor for system effects, k_{sys}, can contain two distinct components. One is the factor arising from the material characteristics, particularly its variability. Ideally, this factor should lead to $k_{sys} = 1.0$ for materials with no variability. The second component is due to idealisations used in the structural analysis. The magnitude of this component depends on the idealisation used in the structural analysis; theoretically this should give a factor $k_{sys} = 1.0$ in the case where the structural analysis procedure used is 'exact'.

An illustration of the difference between these two components is given in Table 8 which shows system factors for a floor system computed by Foschi *et al.* (1989) and the corresponding values in the Australian code AS 1720.1. It is seen that in one case the system factor increases with the number of floor joists, while in the other it decreases. The reason is that AS 1720.1 considers only the system effect due to material variability, i.e. it assumes that a perfect structural analysis procedure is used, while Foschi *et al.* include both components of the system effect, i.e. they assume that idealised assumptions are being used in the structural analysis. Thus, in presenting system factors, it is important to define the assumptions to be used in the associated structural analysis of the system.

5.5 BUCKLING

There appears to be general agreement to include the effects of buckling through the use of a modification factor applied directly to the characteristic strength capacity. The modification factor is a function of material parameters, loading parameters and slenderness. Several definitions have been used for slenderness. These include:

- a definition which leads to $\lambda = L/d$ in the case of rectangular columns,
- a definition in which $\lambda = L/r$ for columns,
- a definition in which $\lambda = \sqrt{R_{squash}/R_{crit}}$

where:

λ = slenderness coefficient
L = column length
d = column depth
r = radius of gyration
R_{squash} = squash load
R_{crit} = critical elastic load.

The first definition is familiar to timber engineers, the second is used in steel codes, and the third is useful for normalising information. For all materials and all types of elements, the third definition leads to the following limiting values of k_{INST}, the modification factor for instability:

for the case of $R_{ult} = R_{squash}$,

$$k_{INST} = 1.00 \tag{35}$$

and for the case $R_{ult} = R_{crit}$,

$$k_{INST} = 1/\lambda^2 \tag{36}$$

where:

k_{INST} = R_{ult}/R_{squash}
R_{ult} = denotes the ultimate load capacity.

Thus, for all materials and types of elements, this definition of slenderness coefficient, λ, leads to an identical plot of the squash load and critical elastic load.

Many buckling strength equations are written as functions of R_{squash} and R_{crit}, and lead to Eqs (35) and (36) as $\lambda \rightarrow 0$ and $\lambda \rightarrow \infty$ respectively.

TABLE 7. Specified Partial Safety Factors for the Danish Code DS 413 (After Danish Standards Institute 1982)

Type of structure	γ_m		
	Low safety class*	Normal safety class	High safety class†
Glulam and other factory-made structures subject to impartial quality control	1.23	1.35	1.50
Other structures	1.36	1.50	1.67

* Failure involves negligible risk to life and small social consequences.
† Failure will involve great risk to life or will have considerable social consequences.

Table 8. Comparison of System Factors for a Floor

Number of floor joists	System factor k_{sys}	
	Foschi *et al.* (1989)	AS 1720.1 (SAA 1988)
2	–	1.11
5	1.58	1.20
10	1.44	1.25
20	1.37	–

TABLE 9. Structural Importance Multiplier for Wind Loads in AS 1170.2 (Standards Australia 1989b)

Class of structure	Structural importance multiplier
Structures which have special post-disaster functions, e.g. hospitals and communications buildings	1.20
Normal structures	1.00
Structures presenting a low degree of hazard to life and other property in the case of failure, e.g. isolated towers in wooded areas, farm buildings	0.80
Structures of temporary nature and which are to be used for less than 6 months	0.65

TABLE 10. Proposed Strength Grouped Properties for Stress-graded Timber in ISO/DIS 8972 (International Standards Organisation 1988)

Strength class	Bending strength (MPa)	Tension strength (MPa)	Modulus of elasticity (GPa)
T75	75	48	19.0
T60	60	38	15.0
T48	48	30	12.0
T38	38	24	9.5
T30	30	19	8.5
T24	24	15	7.5
T19	19	12	6.0
T15	15	9.5	5.4
T12	12	7.6	4.8
T10	9.5	4.8	4.2
T8	7.5	3.8	3.4
T6	6.0	3.0	3.0
T5	4.8	2.4	2.4

TABLE 11. Strength Grouped Properties for Stress-graded Timber for Eurocode 5 (European Committee for Standardisation 1990a)

Strength class	Bending strength (MPa)	Tension strength (MPa)	Modulus of elasticity (GPa)
C60–22E	60	36	22
C48–20E	48	29	20
C37–14E	37	22	14
C30–15E	30	18	15
C30–12E	30	18	12
C24–11E	24	14	11
C21–13E	21	13	13
C21–10E	21	13	10
C18–9E	18	11	9
C15–11E	15	9	11
C15–8E	15	9	8
C13–7E	13	8	7

5.6 FIRE RESISTANCE

The general trend is to use the following format for assessing the fire resistance of timber structures:

(a) for assessing the strength of solid timber or glulam members, the concept of sacrificial timber is used;
(b) for connector systems, ratings are used which have been obtained from furnace tests and are tabulated in LSD codes; and
(c) for composite floor and wall systems, use is made of ratings of components obtained from furnace tests and tabulated in laboratory reports, combined with code rules for assessing the rating when several components are combined.

Some examples of these procedures are given in AS 1720.4 (Standards Australia 1990), Underwriters Laboratories of Canada (1986) and CEI–BOIS (1983).

5.7 IMPORTANCE FACTORS

Partial safety factors for importance are usually related only to loads. An example of this is given in the Canadian code CAN/CSA–086. Another example, taken from the Australian wind loading code AS 1170.2 (Standards Australia 1989b), is shown in Table 9.

However, there are also some examples of importance multipliers that are applied to resistance. This frequently occurs in codes for the design of farm structures (e.g. Standards Association of Australia 1986b).

Another example, noted previously in Table 7, is taken from the Danish code DS 413. Finally it is of interest to note an example from Australia in which importance is linked with quality control; a new recommendation under consideration will require that design stresses for timber that have not been checked by in-grade structural-size testing, be reduced by the equivalence of one stress grade when applied to critical structural elements; here a critical structural element will be defined as one for which failure will lead to loss of life, loss of more than 20 m^2 of floor or loss of more than 30 m^2 of roof.

5.8 STRENGTH GROUPING

Strength grouping is not a structurally efficient method for using timber, but it can be very useful for promoting the utilisation of lesser known species, for marketing and for technology transfer. As an example, the Australian system has been applied to some 2000 species worldwide and in this form it has been used by United Nations agencies to apply a limited set of bridge designs to a large number of countries (Berni *et al.* 1979, Bolza and Keating 1972, Keating and Bolza 1982, UNIDO 1985).

Many LSD codes include some form of grouping. Extracts from two examples are shown in Tables 10 and 11. The ISO example given in Table 10 is intended to have universal and timeless application; the steps between strength classes are in an unbiased geometric progression. The Eurocode 5 example given in Table 11 is obviously targeted at making best use of the timber currently available in Europe; presumably it will need to be changed as market conditions change.

6. Recent Technology

The extensive application of in-grade testing over the past two decades has revealed material characteristics that were unknown at the time of formulating the earlier WSD codes. Some of these matters will have to be included in drafting future LSD codes.

TABLE 12. Proposed Performance Criteria for Timber Structures (after Leicester and Barnacle 1990)

Performance rating	Time to failure* (years)		
	For 50% of beams	For 5% of beams or 50% of joints	For 5% of joints
2	10	5	2
3	50	20	10
4	100+	80	50
5	100+	100+	100+

* For structures loaded to the maximum permissible load according to permissible stress codes.

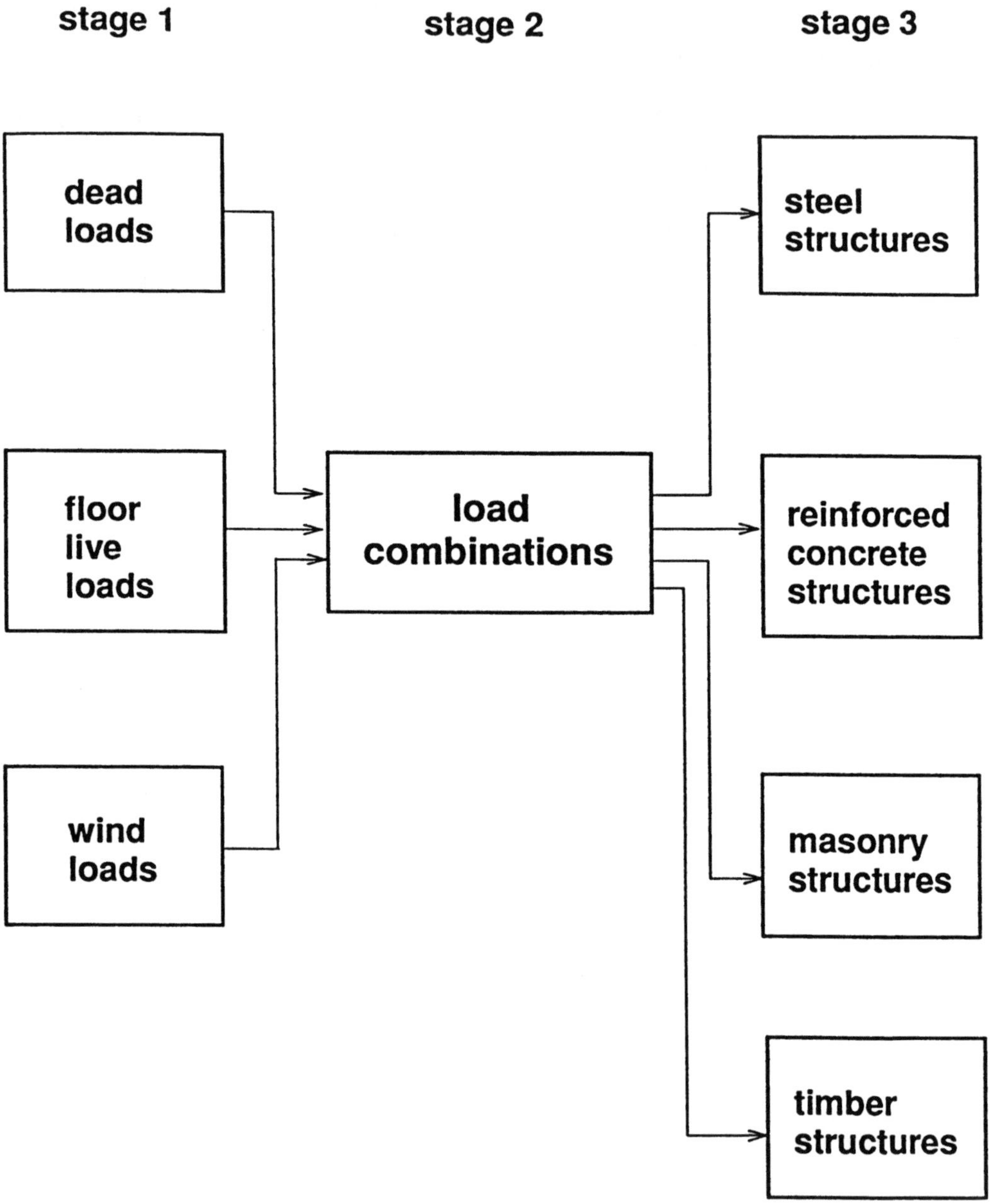

Figure 9. **Sequence of Operations for Coordinating the Development of a Suite of Limit States Design Codes.**

The in-service behaviour of timber is directly influenced by the interactions between environmental conditions and internal stresses (Fridley *et al.* 1990). Thus predictions of both the strength and deformation characteristics of timber are dependent on:

- initial moisture content of the timber,
- eventual in-service moisture content of the timber, and
- fluctuations in the humidity of the environment.

None of the current timber engineering codes, either WSD or LSD codes, take all three of these factors into consideration. The draft LSD code NZS 3603.1 (Standards Association of New Zealand 1991) has an interesting format for including the first two factors.

The long-term interaction of strength and creep that manifests itself in creep buckling behaviour is considered in AS 1720.1 (Standards Australia 1988), but does not appear to be included in any other current code.

It has been fairly well documented that the strength of solid timber members varies with the member length, loading configuration and member depth (e.g. Madsen and Buchanan 1986, Leicester 1985b). Madsen and Buchanan have proposed a format for including these effects in design codes. Ironically, the only factor currently mentioned in codes is the depth effect, which is not a material constant but rather a function of the grading process used; in fact it could be removed if desired.

The grade factor has a marked influence on the characteristics of structural lumber with regard to properties such as tension to bending strength ratio, long duration strength, strength under fire conditions and effect of moisture content. The lower grades tend to be influenced by the characteristics of knots, whereas the higher grades tend to behave like clear wood. There is little information on the effects of kino veins and other defects that control the strength of hardwood lumber. With a few minor exceptions, there is no mention of grade factor effects in current codes.

Finally comment should be made on the matter of durability. In current codes, the vigorous methodology of reliability is abandoned when the design rules are formulated for exposed structures. As a first step towards introducing a formal reliability procedure, a prediction method has been proposed by Leicester and Barnacle (1990) for durability performance as defined in Table 12. An alternative procedure of this type has been given by Masatoshi Sato (1990).

7. Strategies

7.1 CODE DEVELOPMENT

In the development of a suite of reliability-based codes, many independent committees are involved. For the particular case of limit states codes, one method to avoid confusion and conflict is to follow the sequence of operations illustrated schematically in Fig. 9. The sequence is staged as follows:

Stage 1 For each type of load, a code giving characteristic values is developed. Statistical definitions for the characteristic loads are given.

Stage 2 Using the characteristic loads, sets of design load combinations are selected for both ultimate and serviceability limit states. The criteria used for assessing proposed load combinations is to obtain consistency in the reliability index over a wide range of practical situations.

Stage 3 Design codes for various materials are calibrated for use with the specified load combinations.

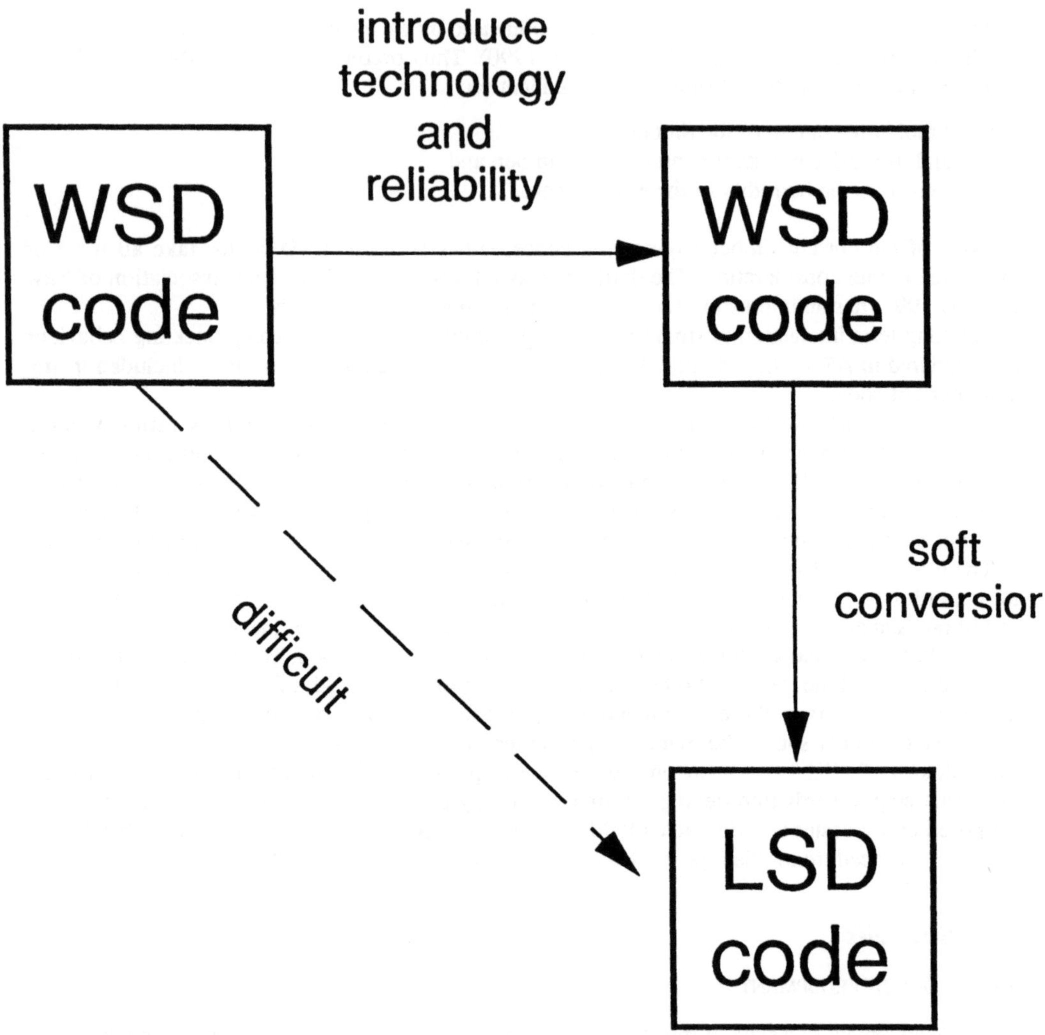

Figure 10. Sequence of Code Developments.

Another useful development sequence is illustrated in Fig. 10. Experience has indicated that if in the conversion to LSD codes there is simultaneously introduced radical new technology and new format, the situation creates confusion to the code user. A simpler transition is obtained if new technology and reliability concepts are first introduced into an interim WSD code and then this is later translated to an LSD code through a soft conversion procedure. This is particularly useful for countries in which there is a policy to accept simultaneously the use of both WSD and LSD codes, because in this way the two codes are coordinated.

The introduction of reliability concepts is sometimes difficult if the membership of a code committee is largely unfamiliar with this technology; in such a case the code committee may in fact even adopt a hostile attitude to the use of reliability methodology. In such instances it is advisable not to attempt to use reliability to formulate the design rules, but rather to use it initially to display the consequences, in reliability terms, of design rules developed from other considerations. Hopefully, as experience is gained, the committee will eventually see the advantages to be gained in using reliability as a yardstick for evaluating design rules and for settling disputes.

7.2 LEGAL CONSIDERATIONS

The introduction of reliability concepts invariably leads to new questions on code formulations which may have legal implications. In particular, a decision must be made as to whether the situation illustrated in Fig. 4 is acceptable. Specifically the decision to be made is whether the reliability index should be the same for all structural materials. In more general terms, decisions have to be made as to whether reliability indices should vary with types of loads and types of buildings.

It is of interest to note that the system effects of parallel support systems, such as floors, have been readily introduced into design codes. These systems tend to enhance the design capacity of individual elements. On the other hand, design codes have been conspicuously silent in ignoring the systems effects which tend to reduce the design capacity of single elements. This includes all series systems, such as trusses. This is yet another matter with legal implications that needs to be resolved.

These difficulties are compounded by the fact that the concept of a target probability of failure, particularly where loss of life is involved, is not acceptable in a legal context. This creates problems in recording and discussing code developments, particularly developments that relate to the concept of differential levels of reliability. The use of the term 'reliability index' rather than 'probability of failure' may be effective in mitigating the problem, but to the author's knowledge, has not yet been tested in a court action.

Finally, it should be noted that with respect the serviceability limit states, the concept of target failure rates would appear to be quite acceptable. For these limit states, failures are often observed and are the basis of frequent litigations. The one difficulty appears to be that litigations often last for many years because of the lack of suitable definitions of failure in the case of serviceability limit states. To avoid this problem, Leicester and Pham (1987) have proposed that a standard be drafted in which serviceability limit states are defined in a manner that enables failure to be checked by making direct measurements in-situ of building parameters in the event of a dispute.

7.3 GLOBAL CONSIDERATIONS

Considerable benefits could be obtained through the general acceptance of a suite of global LSD codes developed by the International Standards Organisation or some other similar body. Examples of such benefits would include the facilitating of trade and technology transfer.

Unfortunately, the difficulties experienced in obtaining agreement for developing a suite of such codes within a single country would tend to indicate that it could be a long process to obtain agreement on global standards. In all probability, this will be obtained through an evolutionary process in which codes for small regions will coalesce to form codes for larger regions which will then eventually coalesce to form a suite of codes that is accepted for global application. One advantage of this process is that it will incur the involvement of a large number of motivated technologists.

8. Conclusions

In the development of an LSD code, many options are available with respect to the choice of matters such as format, characteristic values and safety factors. In this, the use of reliability methodology is useful for quantifying decisions. Additionally, strategies need to be developed to coordinate and progress the many stages required in the development of a suite of LSD codes and their related standards.

9. Acknowledgments

The author is indebted to J.R. Goodman, D.S. Gromala and H.J. Larsen for valuable assistance in the preparation of this paper.

10. References

Berni, C., Bolza, E. and Christensen, F.J. (1979) 'South American timbers – the characteristics, properties and uses of 190 species', CSIRO Division of Building Research, Melbourne, Australia, 229 pp.

Bolza, E. and Keating, W.G. (1972) 'African timbers – properties, uses and characteristics of 700 species', CSIRO Division of Building Research, Melbourne, Australia, 720 pp.

British Standards Institute (1979) BS 5820 'Methods of test for determination of certain physical and mechanical properties of timber in structural sizes', London, UK, 8 pp.

Canadian Standards Association (1989) CAN/CSA–086.1–M89 'Engineering design in wood (limit state design)', Toronto, Canada, December, 234 pp.

CEI–BOIS (1983) 'Technical report relating to the determination of the behaviour of wooden building components exposed to fire', Brussels, Belgium, 91 pp.

Danish Standards Institute (1982) DS 413 'Structural use of timber' (English translation), Copenhagen, Denmark.

Danish Standards Institute (1983) DS 409 'Loads for the design of structures' (English translation), Copenhagen, Denmark.

Ellingwood, B., Galambos, T.V., McGregor, J.G. and Cornell, C.A. (1980) 'Development of a probability based load criterion for American National Standard A58, Building code requirements for minimum design loads in buildings and other structures', NBS Special Publication No. 577, National Bureau of Standards, US Dept of Commerce, Washington DC, USA, June, 222 pp.

European Committee for Standardisation (1989) EN TC 124.202 'Structural timber: the determination of characteristic values of mechanical properties and density of timber' (Draft), Brussels, Belgium.

European Committee for Standardisation (1990a) CEN TC 250/505.12 Eurocode 5 'Design of timber structures: Part 1. General rules and rules for buildings' (Draft), Brussels, Belgium.

European Committee for Standardisation (1990b) EN TC 124.202 'Structural timber – strength classes' (Draft), Brussels, Belgium, 6 pp.

Foschi, R.O., Folz, B.R. and Yao, F.Z. (1989) 'Reliability-based design of wood structures', Report No. 34, Structural Research Series, Dept of Civil Engineering, University of British Columbia, Vancouver, Canada.

Fridley, K.J., Tang, R.C. and Soltis, L.A. (1990) 'Effect of cyclic relative humidity on the load duration behaviour of structural lumber', Proc. 1990 International Timber Engineering Conference, Tokyo, Japan, October, Vol. 2, 407–415.

Goodman, J.R. (1990) 'Reliability-based design for engineered wood construction: update and status of US progress', Proc. 1990 International Timber Engineering Conference, Tokyo, Japan, October, Vol. 1, 6–11.

Gromala, D.S., Sharp, D.J., Pollock, D.G. and Goodman, J.R. (1990) 'Load and resistance factor design for wood: the new US wood design specification', Proc. 1990 International Timber Engineering Conference, Tokyo, Japan, October, Vol. 1, 311–318.

International Standards Organisation (1988) ISO/DIS 8972 'Solid timber-strength grouping', 3 pp.

International Standards Organisation (1990a) ISO 8969 'Timber structures – testing of unilateral punched metal plate fasteners and joints', Geneva, Switzerland, December, 9 pp.

International Standards Organisation (1990b) 'Nails and other dowel type fasteners: determination of bending strength', (Draft N 140), 10 pp.

Keating, W.G. and Bolza, E. (1982) 'Characteristics, properties and uses of timbers, Vol. 1: South-East Asia, Northern Australia and the Pacific', Inkata Press, Melbourne, 390 pp.

Larsen, H.J. (1984) 'Harmonisation of the European timber codes'. Proc. Pacific Timber Engineering Conference, Auckland, New Zealand, May, Vol. 3, 915–922.

Leicester, R.H. (1984) 'Closed form solutions for cost-optimised reliability', Proc. IUTAM Symposium on Probabilistic Methods in the Mechanics of Solids and Structures, Stockholm, 19–21 June, 407–414.

Leicester, R.H. (1985a) 'Computation of a safety index', Civil Engineering Transactions, Institution of Engineers, Australia, Special Issue on Risk, Reliability and Limit States Design, No. CE27, 54–60.

Leicester, R.H. (1985b) 'Configuration factors for the bending strength of timber', Proc. Symposium on Forest Products Research International, Pretoria, South Africa, 22–26 April, Vol. 4, Paper No. 3–8, 13 pp.

Leicester, R.H. (1986a) 'Confidence in estimates of characteristic values', Proc. 19th Conference of CIB-W18, Firenze, Italy, September, 18 pp.

Leicester, R.H. (1986b) 'Load factors for proof and prototype testing', Proc. 19th Conference of CIB–W18B, Firenze, Italy, September, 22 pages.

Leicester, R.H. (1987) 'Load factors for proof and prototype testing', Proc. 1st National Structural Conference, Institution of Engineers, Australia, Melbourne, August, Vol. 2, 546–550.

Leicester, R.H. (1990) 'On developing an Australian limit states codes', Proc. 1990 International Timber Engineering Conference, Tokyo, Japan, October, Vol. 1, 12–20.

Leicester, R.H. and Barnacle, J.E. (1990) 'Durability – a structural approach', Proc. 23rd Forest Products Research Conference, November, Vol. 1, Paper No. 2/4.1.

Leicester, R.H. and Beresford, F.D. (1977) 'A probabilistic model for serviceability specifications', Proc. 6th Australasian Conference on the Mechanics of Structures and Materials, Christchurch, New Zealand, August, 407– 414.

Leicester, R.H. and Pham, L. (1987) 'Serviceability limits', Proc. 1st National Structural Conference, Institution of Engineers, Australia, Melbourne, August, Vol. 2, 430–534.

Leicester, R.H., Pham, L., Holmes, J.D. and Bridge, R.Q. (1986) 'Safety of limit states structural design codes', Seminar Proceedings, Institution of Engineers, Australia, Sydney/Melbourne, March, 180 pp.

Madsen, B. and Buchanan, A.H. (1986) 'Size effects in timber explained by a modified weakest link theory', Canadian Journal of Civil Engineering, 13(2), 218–232.

Masatoshi Sato (1990) 'Guideline of designing the service life of wooden buildings'. Proc. 1990 International Timber Engineering Conference, Tokyo, Japan, October, Vol. 3, 760–764.

Pham, L. and Dayeh, R.H. (1986) 'Floor live loads', Proc. 10th Australasian Conference on the Mechanics of Structures and Materials, Adelaide, Australia, Vol. 2, 559–564.

Pham, L. and Leicester, R.H. (1979) 'Structural variability due to the design process', Proc. 3rd International Conference on Applications of Statistics and Probability in Soil and Structural Engineering, Sydney, Australia, Jan.–Feb., 586–600.

Ravindra, M.K. and Galambos, T.V. (1978) 'Load and resistance factor design for steel', Journal of the Structural Division, ASCE, 104(579), 1337–1354.

Reid, S.G. (1989) 'Risk assessment', Research Report No. R 591, School of Civil and Mining Engineering, University of Sydney, Australia, February, 46 pp.

Standards Association of Australia (1986a) DR 83205 'Draft Australian standard for the evaluation of strength and stiffness of graded timber', Sydney, Australia.

Standards Association of Australia (1986b) AS 2867 'Farm structures – general requirements for structural design', Sydney, Australia, 11 pp.

Standards Association of Australia (1988) AS 1720.1 'SAA timber structures code. Part 1: Design methods', Sydney, Australia, 85 pp.

Standards Australia (1989a) AS 1170.1 'SAA loading code. Part 1: Dead and live loads and load combinations', Sydney, Australia, 29 pp.

Standards Australia (1989b) AS 1170.2 'SAA loading code. Part 2: Wind loads', Sydney, Australia, 96 pp.

Standards Australia (1990) AS 1720.4 'Timber structures. Part 4: Fire resistance of structural timber members', Sydney, Australia, 8 pp.

Standards Association of New Zealand (1991) NZS 3603.1 'Code of practice for timber design' (Draft), Wellington, New Zealand, 63 pp.

Turkstra, C.S. (1972) 'Theory of structural design decisions', Solid Mechanics Study No. 2, University of Waterloo, Waterloo, Canada.

Underwriters Laboratories of Canada (1986) 'List of equipment and materials. Vol. II: Building construction', Ontario, Canada, September, 709 pp.

UNIDO (1985) UNIDO/10/R.162 'Prefabricated modular wooden bridges', Vienna, Austria, 260 pp.

Walford, G.B. (1989) 'Conversion of the NZ timber design code to LSD format', Proc. 2nd Pacific Timber Engineering Conference, Auckland, New Zealand, 305–308.

SAFETY FORMAT OF EUROCODE 5

HANS JØRGEN LARSEN
Danish Building Research Institute
Postbox 119
DK-2970 Hoersholm

1. BACKGROUND

The European Community has agreed to remove internal barriers to the free movement of goods and services. One of the instruments to obtain this is a system of common European building codes: Eurocodes; for timber structures "Eurocode 5, Design of Timber Structures". A first draft was published by the Commission of the European Communities in 1987 (Report EUR 9887) for comments. A final version is expected to be published 1992.

The general design basis is the same in all Eurocodes and is given in a common chapter. To this material-dependent rules are added for timber structures, for example, rules for load duration effects. In the following the safety format for timber structures is described briefly. It is based on an unpublished redraft of Eurocode 5 (April 1991).

2. LIMIT STATES

2.1 GENERAL

The Eurocodes are limit state codes, which simply means that the requirements concerning structural reliability are linked to clearly defined limit states, i.e. states beyond which a specific performance criterion is infringed. Two limit state categories are treated:

Ultimate limit states corresponding to collapse or other states which may endanger the safety of people or result in considerable financial losses.

Serviceability limit states corresponding to states in which service criteria are no longer met, e.g. due to excessive deformations or vibrations.

J. Bodig (ed.), Reliability-Based Design of Engineered Wood Structures, 125–128.

2.2 ULTIMATE LIMIT STATES

2.2.1. Safety Format: Partial Coefficient Method. A partial coefficient method is used. This represents one of the simplest systems possible belonging to the same family of approach as the method with allowable stresses.

In the simplest case, where the load side and the material side can be separated, the Eurocodes require that

$$S_d \leq R_d \tag{1}$$

where:

S_d = design action effect (bending moment, shear force ...) calculated from the following combination of the characteristic values of permanent action (G) and variable actions (Q):

$$S_d = S(\gamma_G G + \gamma_Q(Q_1 + \sum_{i>1} \psi_i Q_i)) \tag{2}$$

R_d = design resistance calculated from the design material parameters X_d (bending strength, tensile strength, ..., moduli of elasticity, ...):

$$X_d = k_{mod} X_k/\gamma_M \tag{3}$$

where:

γ_G, γ_Q = load factors

ψ = combination factor taking into account the reduced possibility at any given time of more than one action having its full characteristic value

k_{mod} = modification factor taking into account the influence of moisture content and load duration

X_k = characteristic material parameter

γ_M = material factor

The main difference between the Eurocodes and, for example, North American practice is that the Eurocodes are based on factored material properties, not factored resistance.

2.2.2 Characteristic Values of Actions. For permanent action the mean value is normally used. If the variability is large, two characteristic values are used, an upper and a lower one. For variable actions the characteristic value corresponds in principle to a 50-year period.

2.2.3 Characteristic Values for Materials. These are defined as the 5-percentiles, based on short-term tests with prescribed load configurations.

It should be noted that Eurocode 5 requires that a strength reducing grade determining defect shall be placed in the most highly stressed zone as opposed to, for example, North American practice where random selection is used. This leads to lower European values.

2.2.4 Partial Coefficients. It is the responsibility of the national authorities to prescribe the partial coefficients. The Eurocodes, however, give a recommended set of values. This set is not based on a rational basis, as for example safety index calculations. Initially, a clear description was given of the different uncertainties to be covered by the load factors and the material factors, but it was later deleted because it was not in accordance with reality. This really does not matter. What count are the global level of safety (described e.g. by $\gamma_Q\gamma_M$) and the ratio γ_Q/γ_G.

The load factors are the result of compromises in prestandardization bodies like the Comité Euro-International du Beton (CEB). The material factors are subsequently determined by calibration to existing design practice. The ratio γ_Q/γ_G is rather low, about 1.1. A ratio of about 1.3 would have ensured a more uniform reliability level.

2.2.5 Load Factors. Table 1 gives the chosen load factors. γ

2.2.6 Combination Factor. Tentative values for the combination factors are

- for imposed loads in dwellings 0.5
- other imposed loads, snow, wind 0.7

2.2.7 Modification Factors. Table 2 gives the the chosen modification factors for load duration and moisture content. They are based on existing practice and guided by recent research results, especially from Canada, USA and the Scandinavian countries.

2.2.8 Material Factors. A material factor, γ_M, of

$$\gamma_M = 1.3 \tag{4}$$

is proposed. This value has been estimated on the basis of a calibration taking the existing codes as benchmark: Generally, the safety factors in the existing codes are about 2.5 (e.g. a long-term factor of about 0.6 or 9/16 and a overload factor of about 1.33). The proposed values give a global safety of 2.70 for a structure with dead load 2.60 for one variable long-term load, and 2.30 for a structure with medium-term load (e.g. snow).

2.3 SERVICEABILITY LIMIT STATES

The most important serviceability limit states for timber structures relate to deflections and vibrations. The design action effect is calculated as for ultimate limit states. Normally, however, all γ-values are taken as unity and lower ψ-values are used.

In the calculation of deflections, the values calculated by using the short-term modulus of elasticity for moisture class 1 (moisture content about 12%) shall be multiplied by $(1 + k_{creep})$, where the creep factor, k_{creep}, is given in Table 3.

Table 1. Partial Safety Factors for Actions.

	Permanent Action, γ_G	Variable Action, γ_Q
Favorable effect	1.0	1.5[1)]
Unfavorable effect	1.35[1)]	1.5[1)]

[1] For one-storey buildings with moderate spans that are only occasionally occupied 10% lower values are used.

Table 2. Modifiation Factors , k_{mod}.

Service Class	1	2	3
Duration of load	< 12%[1)]	12%-18%[1)]	> 18%[1)]
Permanent (more than 10 years)[2)]	0.65	0.65	0.50
Long-term (6 months - 10 years)[2)]	0.75	0.75	0.60
Medium-term (1 week - 6 months)[2)]	0.85	0.85	0.70
Short-term (less than 1 week)[2)]	0.95	0.95	0.80
Instantaneous	1.2	1.2	1.0

[1)] Typical moisture content in softwood
[2)] Order of duration

Table 3. Creep Factor k_{creep} for Different Load Durations.

Service Class	1 and 2 (< 18%)	3
Permanent + long-term	0.5	1.0
Medium-term	0.25	0.5
Short-term	0	0

If members are installed at or near fibre saturation point and are likely to dry out in service, then the values of k_{creep} should be doubled.

LIFETIME OF WOOD AS RELATED TO STRENGTH DISTRIBUTION

LAUGE FUGLSANG NIELSEN
Building Materials Laboratory
Technical University of Denmark
DK-2800 Lyngby, Denmark

1. Introduction

Any realistic reliability study of structures requires information on the reliability of the building materials being used. A reliability analysis of a wood structure, for example, cannot be made without knowledge of the strength and lifetime distributions of the structural wood material. Normally, purely empirical distributions are chosen for each event without any guaranty that this procedure will not introduce inconsistency which may influence in some *un-known* way the results of a reliability analysis.

These considerations encompass the topic of the present paper. It is demonstrated that the two distributions considered are correlated, meaning that lifetime distribution can be predicted from strength distribution. An algorithm is developed for this purpose - and a discussion is made on how to extract a maximum of lifetime information from experimental data. The following symbols are used in the paper:

σ_l	Theoretical strength
σ_{cr}	Real (short time) strength
$FL = \sigma_{cr}/\sigma_l$	Strength level (Materials quality)
σ	Load
$SL = \sigma/\sigma_{cr}$	Load level
t	Time
b	Creep power
τ	Relaxation (creep doubling) time
t_{cat}	Time to catastrophic failure

2. Analysis

Wood is a damaged viscoelastic material (DVM) which becomes weaker as damages approach their critical sizes. This materials concept (DVM-concept) has been developed by the author in recent years to explain and predict successfully a number of wood mechanical properties such as lifetime and residual strength under static and varying load (e.g. author 1978; 1979; 1991a; 1991b; author and Kousholt 1980). The DVM-expressions relevant for the analysis of deadload lifetime are summarized in Section 2.1. The distribution

J. Bodig (ed.), Reliability-Based Design of Engineered Wood Structures, 129–138.

of lifetime is then studied by "experiments" where a group of structural wood members is subjected to constant loads chosen from the lower end of the associated strength distribution.

Methods of estimating strength levels and creep parameters of wood are suggested by the author in (1991a; 1991b). The strength level, for example, is estimated by the following (simplified) expression:

$$FL = \sqrt{1 - e^{-d/R}} \quad ; \quad (\approx \sqrt{d/R} \quad at \; R/d > 10) \tag{1}$$

where R is defect radius (like crack or knot for example) and d is a characteristic micro-structural dimension of *"model wood"* (virgin clear wood with no "defects" like pits and rays for example). A quantity of $d \approx 0.3mm$ is suggested as square root mean of length and diameter of a wood fiber. The reference (theoretical) strength, σ_l, is strength of a bundle of model wood fibers. Structural wood have strength levels in the range of FL < 0.2–0.3 and clear wood in the range of 0.2–0.3 < FL < 0.8 (where R = d). Strength ratios of FL > 0.8 can only be obtained by "healing" the structure of clear wood such that any defect radius is smaller than d.

The creep parameters, τ and b, define the *local creep* behavior of wood at the crack vicinities by the the well-known Power-Law creep function formulated as follows by the present author in (1984):

$$C(t) = 1 + (t/\tau)^b \tag{2}$$

A creep power of b ≈ 0.25 and a relaxation time of $\tau \approx 10^{1 \pm 1}$ days (dry wood in bending at 20 °C) were suggested by the author in (1991a, 1991b). The relaxation time is highly sensitive to climatic conditions (creep power is not). The actual value of τ, however, is of no interest in the present context as the analysis is based on non-dimensional time (t/τ).

The formulation of the DVM-theory with non-dimensional time is very convenient in lifetime analysis. Lifetime is predicted to be proportional to relaxation time – and the influence on lifetime of temperature and moisture is easily considered (author e.g. 1991a). In general the DVM-theory operates with non-dimensional quantities (strength level, load level, damage ratio) which makes it independent of orthotropy and mode of loading when wood is considered. In this way the DVM-approach also qualifies as a so-called damage accumulation theory.

2.1 LIFETIME EXPRESSIONS

Deadload lifetime, t_{cat}, of a damaged viscoelastic material like wood is predicted by the following relation (author 1991b):

$$\frac{t_{cat}}{\tau} = (A - 1)^{1/b} + \frac{Bq}{C} H(\mu) \tag{3}$$

where the parameters involved have the meanings expressed by Equations (4) through (10). The limit quantities in parenthesis apply at low strength levels (FL → 0).

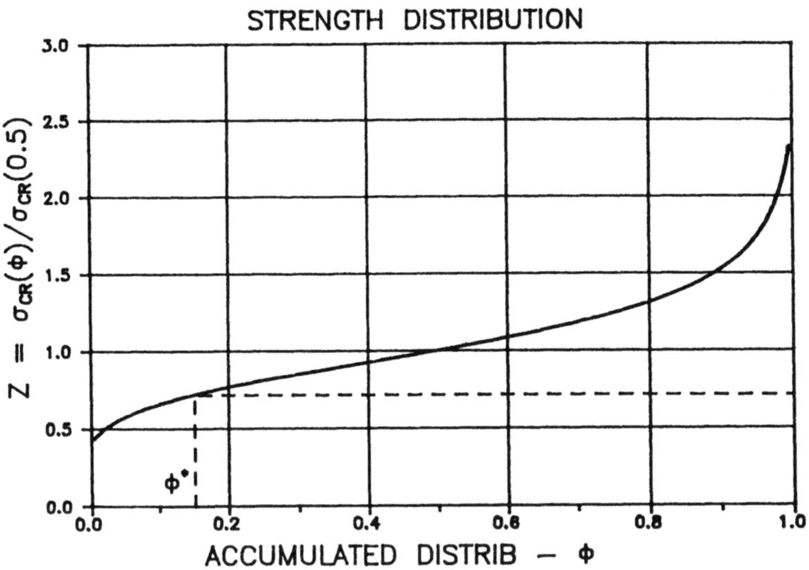

Figure 1. **Strength Distribution According to Eq. 11, with (a,d) = (0.01,0.20).**

$$q = [(1+b)(2+b)/2]^{1/b} \tag{4}$$

$$A = \frac{\ln(\cos(\pi FL/2))}{\ln(\cos(\pi FL*SL/2))} \quad ; \qquad (A \rightarrow SL^{-2}) \tag{5}$$

$$B = \beta + (1-\beta)*\cos(\pi*FL*SL/2) \quad ; \quad (\beta = (4-b)/5) \quad ; \quad (B \rightarrow 1) \tag{6}$$

$$C = 1 - \cos(\pi*FL*SL/2) \quad ; \qquad (C \rightarrow (\pi*FL*SL)^2/8) \tag{7}$$

$$H(\mu) = \int_0^\mu \frac{x^{1/b}}{1+x}\,dx \tag{8}$$

$$\mu = \begin{cases} A*B-1 & \text{when } A*B \geq 1 \\ 0 & \text{when } A*B < 1 \end{cases} \tag{9}$$

In general H(μ) can only be determined numerically. For special creep powers, however, the integral involved can be developed analytically. Some useful examples are given in a reference by the author (1991b). For b = 1/4, which is used in the present study we obtain:

$$H(\mu) = \frac{\mu^4}{4} - \frac{\mu^3}{3} + \frac{\mu^2}{2} - \mu + \ln(\mu+1) \qquad (b = 1/4) \tag{10}$$

2.2 STRENGTH AND LOAD DISTRIBUTIONS

Now consider a group of wood members defined by a characteristic strength distribution which are fitted non-dimensionally with respect to the 50-percentile strength, $\sigma_{cr}(0.5)$. A number of mathematical expressions are available for this purpose. A very simple one is suggested for use, namely:

$$Z = Z(\Phi) = \frac{\sigma_{cr}(\Phi)}{\sigma_{cr}(0.5)} = \left(\frac{a+\Phi}{1+a-\Phi}\right)^d \tag{11}$$

Z = Z(Φ) is the Φ-percentile (relative)strength, while a and d are characteristic distribution parameters. The example shown in Figure 1 with (a,d) = (0.01,0.20) is a reasonable order of magnitude estimate of a strength distribution for structural wood. (The parameter Φ will subsequently also be referred to as "member number" in an array where members of the group are ranked from Φ = 0 to Φ = 1 by increasing strength).

The strength level of each member (number Φ in Figure 1) is expressed by Z as follows where the 50-percentile strength level is given by FL(0.5) = $\sigma_{cr}(0.5)/\sigma_t$.

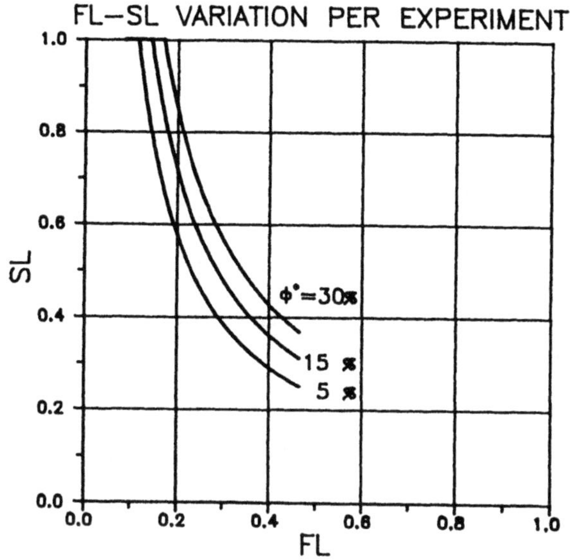

Figure 2. Variations in Strength and Load Levels According to Eqs. 12 and 14, Respectively. FL(0.5) = 0.2. Strength Distribution from Figure 1.

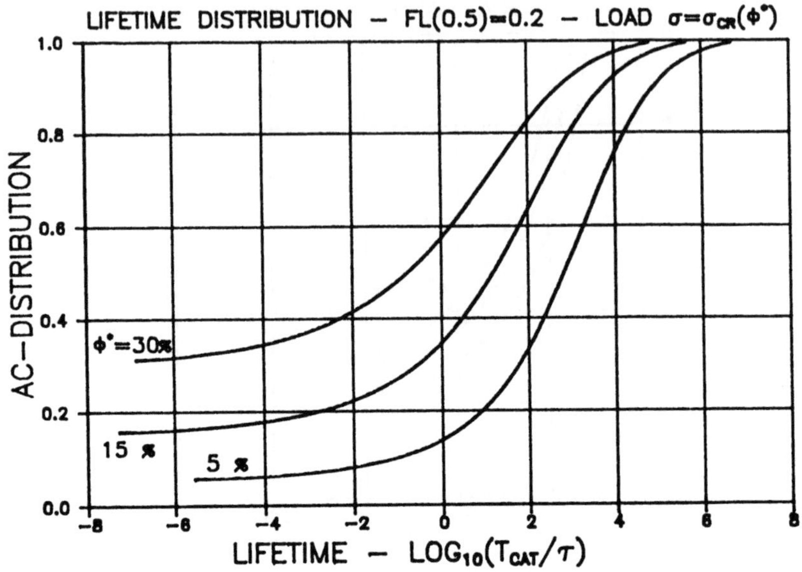

Figure 3. **Lifetime Distributions Associated with the Strength Distribution Given in Figure 1 and Loads as Indicated.**

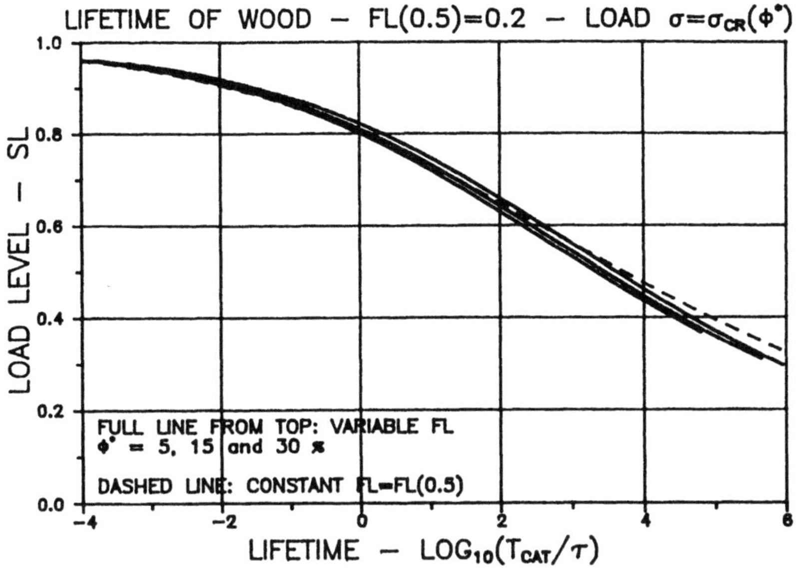

Figure 4. Lifetime of Wood Recorded at Constant Loads as Indicated. Wood Quality is FL(0.5) = 0.2 with Strength Distribution Given in Figure 1.

$$FL(\Phi) = FL(0.5)*Z(\Phi) \qquad (12)$$

It has been indicated previously that normal wood has a strength level which cannot exceed 0.8. This means that Eq. (11) (or another approximation) and Eq. (12) are tied together by the following condition:

$$Z(1) * FL(0.5) \le 0.8 \quad \rightarrow \quad a \ge [(\frac{0.8}{FL(0.5)})^{1/d} - 1]^{-1} \qquad (13)$$

Each member experiences the following load level when the group is loaded with σ. The latter term in Eq. (14) applies when σ appears in the strength distribution as $\sigma = \sigma_{cr}(\Phi^*)$, (see Fig. 1).

$$SL(\Phi) = \frac{\sigma / \sigma_{cr}(0.5)}{Z(\Phi)} \quad ; \quad (= \frac{Z(\Phi^*)}{Z(\Phi)} \quad at \quad \sigma = \sigma_{cr}(\Phi^*)) \qquad (14)$$

The variations of strength level and load level per "experiment" defined in the following section are predicted by Eqs. (11), (12), and (14). These variations are presented graphically in Fig. 2. A load level of SL > 1 is not possible, meaning that the members with $\Phi \le \Phi^*$ will fail during the process of loading.

2.3 LIFETIME DISTRIBUTION – AN EXAMPLE

A group of wood members with shorttime strength defined by the distribution given in Fig. 1 is used in three lifetime "experiments" with loads given by the 5–, 15–, and 30–percentile strength, i.e. $\sigma = \sigma_{cr}(0.05)$, $\sigma_{cr}(0.15)$, and $\sigma_{cr}(0.30)$, respectively. A creep power of b = 0.25 is estimated as suggested in (author 1991a).

Algorithm: Lifetime for each member (number Φ) of the group is predicted by Eqs. (3) through (10) with *strength level* and *load level* determined by Eqs. (12) and (14), respectively with Φ ranging from 0 to 1 (see Fig. 2).

Lifetime distribution: The lifetime of each member is related to its number (Φ) in Fig. 3 which then represents the distributions of lifetimes obtained in our experiments.

Lifetime: The t_{cat}–SL results are shown by solid lines in Fig. 4. Notice that each graph represents a constant load on a material of varying quality (members with different damage structures). The true lifetime of a certain quality wood is constructed by connecting points of equal FL values between graphs. It is, however, remarkable how qualified a lifetime estimate in our example can be obtained directly from Fig. 4. Apparently the effects of SL becoming lower and FL becoming higher with increasing Φ neutralize each other practically such that lifetime at some average strength level is well described by data from only one experiment. This statement is justified comparing the solid lines of Fig. 4 with the dashed line which is lifetime as predicted at a strength level of FL ≡ FL(0.5). It should be noticed, however, that broader selected wood (than defined by Fig. 1 approximately) will increase the separation between the graphs of Fig. 4 and reduce the quality of the simple lifetime prediction method.

3. Conclusions and Final Remarks

The constant load "experiments" described in Section 2 show that lifetime observations may differ by approximately 4 decades between load levels, SL ≈ 0.9 and SL ≈ 0.6. This order of magnitude was also attained in real experiments (Madsen and Johns 1982a; Hoffmeyer 1990). It seems then justified to suggest that lifetime distribution is in fact a logical consequence of strength distribution. The two distributions can not be considered as non-correllated characteristics of wood.

Each experimental lifetime observation reflects the behavior of an individual and should be respected as such. This feature has been considered in Section 2 from which the following summarized procedure is suggested to obtain a maximum of lifetime information from experiments:

- *Step 1:* Lifetime data from a number of constant load experiments are plotted (as in Fig. 4) against individual load ratios determined by Eq. (14).
- *Step 2:* Each data point is marked with its (relative) strength level according to Eq. (12).
- *Step 3:* Lifetime of wood of quality "FL" is then constructed by connecting data points marked "FL" between graphs.

A simplified version of this procedure is justified by the example in Section 2.3: When wood is better selected (lower d) than given by Fig. 1 as an approximation one may skip Steps 2 and 3 (the "FL-marking" procedure) and consider the lifetime-SL graph (obtained by only one experiment) directly as real lifetime applying to some average quality of the wood material considered at $\Phi > \Phi^*$. However, smaller sections of lifetime information are described in this way as the selection quality increases ($d \rightarrow 0$ in Eq. (11)). The statement made in Section 2.3 that t_{cat}-SL graphs move closer when materials selection becomes narrower complies well with experimental data reported in (Madsen and Johns 1982a; Hoffmeyer 1990) with the latter reference representing the narrower selection.

The well-known Borg Madsen method introduced in (Madsen 1971) of treating experimental lifetime data is consistent with Step 1 of the general procedure described above. The constant quality lifetime graphs ("design curves") developed theoretically in (Madsen and Johns 1982a, 1982b; Johns and Madsen 1982) to illustrate the power of the DVM-method are equivalents to the constant FL-graphs which can be deduced directly from experimental evidence by the three step procedure previously described.

Finally, it should be mentioned that the results presented in this paper are valid in principle for any building material which can be modelled like a damaged viscoelastic material. This feature is explained in more details by the author in (1991a, 1991b).

4. Summary

Lifetime of wood is related to load level as well as wood quality. The load level statement is a matter of course. The wood quality statement has been justified in recent years both theoretically and experimentally. It was shown in this paper that lifetime distribution is itself a quality dependent quantity. An algorithm is developed predicting lifetime distribution directly from strength distribution. Analysis of wood structures must consider distributions of strength and lifetime as correlated quantities. A discussion was presented on how to extract a maximum of relevant lifetime information from experimental data – and it was finally pointed out that conclusions drawn in this paper are valid also when a number of other building materials are considered.

Literature

Hoffmeyer, P. 1990: "Failure of Wood as Influenced by Moisture and Duration of Load". Thesis, Dept. Environmental Science and Forestry, State Univ. of New York, Syracuse, N.Y.

Johns, K. and Madsen, B. 1982: "Duration of Load Effects in Lumber, Part I: A Fracture Mechanics Approach". Can. J. Civ. Eng., 9: 502–514.

Madsen, B. 1971: "Duration of Load Tests for Dry Lumber in Bending". Struct. Res. Series, Report No 3. Civ. Eng. Dept., Univ. Brit. Columbia, Vancouver B.C., Canada.

Madsen, B. and Johns, K. 1982a: "Duration of Load Effects in Lumber, Part II: Experimental Data". Can. J. Civ. Eng., 9: 515–525.

Madsen, B. and Johns, K. 1982b: "Duration of Load Effects in Lumber, Part III: Code Considerations". Can. J. Civ. Eng., 9: 526–536.

Nielsen, L. Fuglsang 1978: "Crack Propagation in Linear-Viscoelastic Materials" (in danish with extended english summary), Bygningsstatiske Meddelelser, 49: 1 – 45.

Nielsen, L. Fuglsang 1979: "Crack Failure of Dead-, Ramp-, and Combined Loaded Viscoelastic Materials", First Int. Conf. on Wood Fracture, Banff, Alberta, Canada, 1978. Proc. Western Forest Products Lab., Univ. Brit. Columbia, Vancouver, Canada: 187 – 200.

Nielsen, L. Fuglsang and Kousholt, K. 1980: "Stress-Strength-Lifetime Relationship for Wood". Wood Science, 12: 162–164.

Nielsen, L. Fuglsang 1984: "Power Law Creep as Related to Relaxation, Elasticity, Damping, Rheological Spectra and Creep Recovery – With Special Reference to Wood". IUFRO-Engineering Group Conference 1984. Xalapa, Mexico. Proc. Techn. Univ. Denmark, Build. Mat. Lab., Copenhagen: 181 – 204.

Nielsen, L. Fuglsang 1991a: "Lifetime and Fatigue of Wood and Other Building Materials Subjected to Static and Repeated Loads". IUFRO-Engineering Group Conference 1990, Saint John, New Brunswick, Fredericton, N.B., Canada. Proc. Techn. Univ. Denmark, Build. Mat. Lab., Copenhagen: 448–482.

Nielsen, L. Fuglsang 1991b: "Lifetime, Residual Strength, and Quality of Wood – and other viscoelastic building materials". Holz als Roh- und Werkstoff (november, in press).

Probabilistic Based Safety Factors in the Netherlands

by
Ad.J.M.Leijten
University of Technology Delft
P.O. Box 5048, 2600 GA Delft,
The Netherlands

Summary
In the early eighties it was decided for the new generation of codes in The Netherlands that for all building materials the same safety concept should be pursued. The Safety Project should provide a set of safety factors factors for the load and resistance factor design procedure introduced by the new generation of codes. Outlined are the procedures used and results for the different building materials including timber. It appears that the safety of timber structures is comparative with steel and concrete structures.

Introduction
In the design of structures a margin of safety is introduced. Usually this margin can be found on the strength side in order to obtain allowable material stresses. After 1970, in many countries, a change was made to the limit state design in which the safety margin was accomondated entirely on the load side, to allow plastic theory. In present decades most codes of practice for structural design have adopted partial safety factors to insure a safety margin for the material strength as well as for the action side. The magnitude of these factors can be based on probabilistic analyses. Below a review is given of the various historic approaches used in the Netherlands.

Code TGB 1955 $\quad \dfrac{R}{\gamma} \geq S \quad$ Resistance factor design

Code TGB 1972 $\quad R \geq \gamma\, S \quad$ Load factor design

Code TGB 1990 $\quad \dfrac{R}{\gamma_r} \geq S^* \quad$ Load and resistance factor design

with

$$S^* = \gamma_g G + \gamma_e \psi_T Q_{e1} + \sum \gamma_e \psi_i Q_{ei} \quad \text{or} \quad \gamma_g G + \sum \gamma_e \psi_i Q_{ei}$$: fundamental combination

J. Bodig (ed.), Reliability-Based Design of Engineered Wood Structures, 139–145.

The assessment of partial material and load factors should be based on a unified bases. The same load models should be used independent of the material under consideration. The following fundamental starting points were formulated:
-calculations should be based on limit state design,
-different safety classes with clearly defined degree of reliability should be introduced for different situations,
-calculation should be probability based which supports the characteristic values and partial safety factors, both for each source of loading and for each material,
-material factors should be independent of load factors, and vice versa,
-rules for load combinations should likewise be statistically supported.

The foregoing considerations led to the so-called "Safety of Structures" project which ended in 1984.

The result is that the new generation codes have prescribed safety factors for the materials and the loads. The regulations in the code are then regarded as leading to structures with an acceptable probability of failure (β - safety index).

As the codes are now part of our building law there is an escape procedure built in for those cases were the codes do not apply. When designers want to calculate elements or complete structures in a different way then prescribed in the codes, or when the codes does not provide any information, they can be obligհed to determine the safety index by means of reliability safety analyses to satisfy the building authorities. In the near future software will be offered which contain probabilistic information about loads to be applied for this design approach. On one hand it provides the designer more freedom on the other hand these calculations are perhaps more difficult than one expect. No experience is yet established and more than one eye is focused an those who will make a first attempt.

For this approach not only load models are an necessary ingredient also information about the properties of the building material is required. How fast materials like steel and concrete will advance in this direction is still uncerain not to speak of timber.

Set up of the project

The purpose of the project was to get indicative values for load, material and load combination factors. As a starting point it was assumed that the existing codes lead to structures which posses an acceptable level of safety in order to make some calibration possible. Therefore a number of simple construction elements like beams,

Table 1: Overview of the structural elements considered.

material	concrete	steel	**timber**		
element	beam	column	joint		
loading	dead load	live load	wind	snow	
limite state	ULS (collapse)	SLS (deflection)	SLS (crack-width)		
type of building	office	dwelling	sports hall	warehouse	theatre

Table 2: Average value of reliability index β (reference periode 50 years)

limit state	structural element	average β steel	concrete	**timber**
ultimate load	floor beam	4.2	4.7	**3.9**
	roof beam	2.7	4.5	**3.1**
	pin-ended column	5.0	6.1	**4.8**
	unbrached column	2.2	2.5	**2.3**
	joint	4.0	4.7	**3.0**
serviceability	floor beam	2.8	-0.6	**3.0**
	roof beam	1.6	1.2	**2.2**

Table 3: Proposals for partial safety factors

	strategy 1: min $\sum(\beta-3.8)^2$ (ULS)			strategy 2: min $\sum(\beta-\beta_{now})^2$		
quantity	ULS	combination factors	SLS	ULS	combination factors	SLS
loads						
self-weight	1.20	-	1.00	1.20	-	1.00
live load	2.00	0.90	0.90	1.70	0.90	0.90
snow load	2.13	0.46	0.46	1.70	0.50	0.50
wind load	2.22	0.50	0.50	1.50	0.50	0.20
materials						
structural steel	1.00	-	1.00	1.10	-	1.10
reinf.steel	1.10	-	1.00	1.30	-	1.00
compr.strength concrete	1.10	-	1.00	1.70	-	1.00
standard timber	**1.30**	-	**1.00**	**1.20**	-	**1.10**
glulam	**1.30**	-	**1.00**	**1.20**	-	**1.10**

columns and joints have been designed with the existing codes for steel, concrete and timber. Afterwards they were subjected to a FOSM (First Order Second Moment) reliability analysis . For all elements, materials and failure mechanisms this procedure led to a value for the safety index β and partial safety factors γ. There appeared to be a consideral scatter in the values (partial factors) associated with one load or material strength when taken as a starting point. In order to formulate useful proposals these factors should be equalised on some bases.

This should eventually lead to:

- Load factors for self weight, floor, snow and wind loads for the ULS (ultimate limit state) and SLS (serviceability limit state),
- Material factors for steel, reinforced concrete and (laminated) timber for ULS and SLS, independent of the loads.

The structural elements considered are presented in Table 1. The mean value of the reliability index β found is given in Table 2. It appears that the existing design practice for the ultimate limit state yield β-values ranging from 2.2 to 6.1. The average value for β is 3.8 with a standard deviation of 1.4. The differences between the materials is not substantial. For end-pined columns and for loading cases with a high proportion of self weight the β-values are high and independent of the material. Low β-values are found where variable loads dominate such as for wind load on unbraced columns. The negative value for the serviceability limit state of concrete beams is intriguing. It indicates that the average crack width and deflection

exeed the standard values. The following reasons were given why in practice for most structures the actually reliability index is greater:

- practical rounding of values in determining the dimensions have not been taken into account,
- only failure governing load cases have been considered in the project,
- the co-operation of various elements as a structural system has not been taken into account,
- hidden reserves of safety have not been considered.

These positive effects can be expected to be similarly present in future structures.

Determination of the partial safety factors

As part of the project a computer program was developed able to calculate the value of the reliability index for a proposed set of partial safety factors for all building elements. This program is based on an approximation because for a precise answer the level II calculations as well as the dimensioning should be repeated. With the aid of this program the average and scatter of the safety level for any given set of partial safety factors can be determined. In this way a bases for judgement for a given set of partial safety factors was created.

Besides this the program is able to select from a certain set of partial factors, which range between certain boundaries, those values which leads to a minimum of scatter for a required safety level. With this option two strategies have been analysed:

1) A choice of safety factors that will lead to a minimum of scatter with reference to one preferred value of β for the ultimate limite state and one value of β for the serviceability limite state. The aimed β-values are 3.8 and 1.7 which are in conformity with the average levels found for the existing design procedures as determined in the project.
2) Such a choice of safety factors that the β-values according to the new codes would differ as little as possible from the historical accepted safety levels, the so-called soft calibration.

Since the program optimizes in a strictly mathematical way without knowledge about the definition of a partial safety factor, the value should be proportional to the uncertainty and its influence, manual corrections were necessary.
Because Level I partial factors are derived from a level II approach as shown below these load and material factors are on longer independent of each other.

Level II: linearization at the design point

$$S^* = \mu(S) - \alpha_S \beta \sigma(S) \qquad S^* = \gamma_s \cdot S_{kar}$$

$$R^* = \mu(S) - \alpha_R \beta \sigma(S) \qquad R^* = \frac{R_{kar}}{\gamma_r}$$

notation for level I calculation

with
α = coefficient of influence
β = reliability index
σ = variation coefficient

$$\gamma_s = \frac{\mu(S)}{S_{kar}} \left[1 - \alpha_S \beta \frac{\sigma(S)}{\mu(S)} \right]$$

$$\gamma_r = \frac{R_{kar}}{\mu(R)} \left[\frac{1}{1 - \alpha_R \beta \frac{\sigma(S)}{\mu(S)}} \right]$$

$$\alpha_R = \frac{\sigma(R)}{\sqrt{\sigma^2(R) + \sigma^2(S)}} \quad \text{and} \quad \alpha_S = \frac{-\sigma(S)}{\sqrt{\sigma^2(R) + \sigma^2(S)}}$$

from which follows that material and load factors are dependent
The results are summarized in Table 3.

Table 4: The imput of the statistical properties of the materials

material	type	mean	cov
steel FE 360	LN	280Mpa	0.08
concrete B 22.5	LN	22 Mpa	0.15
standard timber	W	30 Mpa	0.18
tolerances	LN	nominal	0.04
exentricities e/l	TN	$10^{-3}$to 10^{-2}	0.80
weight	N	nominal	0.05
live load	E x I	1.5 kPa	0.40
wind load	E x I	1.0 kPa	0.20
snow load	E x I	0.8 kPa	0.25

LN = lognormal
N = normal
TN = truncated normal
W = Weibull
E x I= extreme type I

Table 5: Consequences for material costs in [%]

	strategy 1			strategy 2		
ULS	steel	concrete	**timber**	steel	concrete	**timber**
beam (floor)	- 7	- 8	**+33**	- 4	- 2	**+16**
beam (roof)	+ 7	- 5	**+22**	+ 4	- 2	**+16**
pin-ended column	-22	-22	**+ 7**	19	- 8	**-11**
unbrached column	+ 4	+ 4	**+36**	- 2	- 8	**- 1**
SLS						
floor beam	- 7	+ 5	**- 4**	+4	+ 4	**+ 2**
roof beam	- 7	+ 8	**- 7**	0	+ 4	**- 2**

Results for timber structures

The statistical properties of all building materials considered are presented in Table 4 Results show that timber is not to different from the results of steel and concrete. Nethertheless it is remarkeble that according to the dimensioning of Table 5 where β = 3.8 is aimed for all materials and element types, elements made of timber require much more material than according to existing practise. As shown 36% more for the column in case of governing wind loads but also +33% and +22% for floor and roof beams respectively and +8% for the pin-ended column. As a reason for this the authors state that for instance from the calculations for floor beams it follows that the relative contribution of the material strength in the total variation (scatter) due to all influencing stochastic variables is about 88% . Also for other timber construction elements the contribution of the material strength in the variation is considerable.

According to the authors of the end-report this is not so strange as timber is a product of nature and has a considerable variation in material strength. This should also be the explanation for the big difficulty to improve the β-value for timber. Furthermore they question whether it is sensible to aim at the same reliability index for timber as for steel and concrete. However the calculations based on the proposed partial safety factors make clear that there is still room to play around with material related factors like the long duration factor to smooth the change to reliability based ultimate limit state design.

As in the early eighties the long duration strength formulation was not so refined as now (load accumulation models), only two material strength situations were considered, short term and long term strength. It appeared that in most cases the short term loading was decisive only in case of high and long permanent load, for instance in storage rooms, the long duration strength had a little influence. When I tried to study the reports related with this Safety Project, I found some mistakes in the calculations which formed the backgroud for the figures in Table 5. In these calculation the long duration factor was always set 0.56 irrespectively of the loading type under consideration. Now we try to find out to what extend the factors will change when proper values are applied.

Literature

Vrouwenvelder, Siemens, " Probabilistic calibration procedure for the derivation of partial safety factors for the Netherlands building codes", Heron, vol.32, No.:4, 1987
Siemens, Vrouwenvelder," Safety of buildings"(in Dutch), TN)-report BI-84-36, 1984
Vrouwenvelder," Background of the safety factors TGB-1986" (in Dutch), Part of the "Construeren in Hout "symposium 1986, Delfts University Press.

GROUP A: FUNDAMENTALS OF RELIABILITY ASSESSMENT

Organizing Committee Member	:	Hans Larsen (unable to attend)	Denmark
Speaker	:	Henrik Madsen	Denmark
Group Leader	:	Sven Thelandersson	Sweden
Rapporteur	:	Frédéric Rouger	France
Participants	:	Bruce Ellingwood	U.S.A.
	:	Zsolt Kovacs	Hungary
	:	John Zahn	U.S.A.

A1. INTRODUCTION

This group offered the opportunity to exchange ideas between a theoretician of reliability and researchers in timber engineering. As a starting point for the discussion, we reviewed the basic ideas exhibited in the paper presented at the ARW on the "Fundamentals of Reliability Assessments", which were:

- uncertainty sources
- definitions of limit states functions
- numerical methods available for safety index calculations
- sensitivity analysis
- updated reliability analysis
- reliability-based optimal design

It clearly appeared to the group that most of the fundamental reliability analysis tools had been already developed. Therefore, the aim would be mainly to investigate the specificities of timber structures, and identify which techniques are most suitable to perform reliability calculations. This working group topic has strong interactions with other groups' topics. First of all, the evaluation of material resistance (numbers of tests to perform, statistical fitting techniques, goodness-of-fit estimators, ...) has a major influence on reliability calculations. Concurring this aspect, the group placed the emphasis on techniques available to measure model uncertainties with reliability calculations. The other interconnected working group subject is the reliability of systems. Most recent progress in reliability theory has been focused on the evaluation of system performance. Therefore, our task was to prepare the "ingredients" necessary for a system calculation. The group identified seven topics in which the state-of-the-art could be established and research needs clearly expressed.

The first two topics concern basic material behavior. The <u>time-dependent behavior</u> is wood specific and requires adapted techniques to handle. This behavior, influenced by moisture content and temperature, has reliability consequences for both

J. Bodig (ed.), Reliability-Based Design of Engineered Wood Structures, 147–157.

serviceability and ultimate limit states. The spatial-dependent behavior, due to the material constitution, manifests itself in the size effects, the geometry effect of glue laminated beams, and in the local inhomogeneous behavior.

The third topic is a review of reliability bases for design. First of all, the way to formulate objective functions for reliability based optimal design is examined. Once a structure has been identified, the field experience might give additional information for calculations. Beyond design requirements, the serviceability limit states are of major importance for the behavior of timber structures. Besides probabilistic aspects, different deterministic equations are formulated. The limit states equations, the constitutive equations of the material, and the mechanical representation of connections are just a few of the many models that need to be verified through reliability calculations. The identification of behavioral models is, therefore, an important issue within this subject.

To handle these different problems, available tools for reliability analysis are reviewed, and the random field and random process analysis techniques are scrutinized. The use of stochastic finite elements for spatial variations is also discussed; techniques of sensitivity analysis are discussed as a way to measure the robustness of the different assumptions.

The last topic considered in detail is education and communications. Reliability techniques are fairly recent and need to be taught to civil engineers and material scientists. Also discussed is the opportunity to create cooperation among scientists or experts on various materials which will facilitate the transfer of knowledge and may minimize duplication of efforts.

A2. TIME-DEPENDENT BEHAVIOR

The time-dependent behavior is very important for the definition of ultimate strength and deformation characteristics of wood. Variation caused by time-dependent behavior is more significant for wood than for steel and concrete, and requires new developments in stochastic modeling of loads, material properties, and their interactions.

The time-dependent phenomena are conveniently divided into load-dependent phenomena, such as creep, and load-independent phenomena, such as various forms of biological degradation.

To analyze time-dependent behavior, it is necessary to introduce a random process model for the loads since the extreme load value alone is an insufficient time characteristic. Two types of load models appear to be particularly attractive -- the Gaussian process model and the filtered Poisson process model. Wind loading may, as an example, be modelled as occurring according to a Poisson process in time and as a Gaussian process within each short-term interval. Other important loads resulting from snow, normal and extraordinary occupancy, temperature and humidity are also well described by the two process types. Related to the time-dependent material behavior are the occurrence rates, dwell times above certain levels, and their distributions at a given time. These are important functions and need to be available for the two process types.

At present, there is a lack of knowledge related to some important loads of which two shall be mentioned. The magnitude and duration of extraordinary occupancy loads are not well documented. Some field surveys have been made, but the data are rather limited. The same can be stated about storage loads on roofs.

Moisture in a structural element is very important for its long-term behavior. Applying a spectral representation of the humidity variation and determining the filtering effect from diffusion in and out of the wood represents a possible approach. Better knowledge about the humidity variation in a building is needed and the filter characteristics for real wood elements with natural defects should be developed. Diffusion into homogeneous material is easily described once the relevant material properties are available. However, complications arise where knots are present and fiber angles turn perpendicular to the surface, reducing the resistance to moisture transportation. To determine the filtering characteristic for an element with a variable number of randomly placed defects is thus an important task.

Material models for such time-dependent behavior as creep, crack growth, residual strength, and change in stiffness are at the present time to a large degree empirically based. In addition to natural variability, uncertainty related to the models themselves and the estimation of model parameters must be addressed. Stochastic models have been developed for other materials, e.g. creep of concrete and crack growth in metals. These models should be considered as starting points for wood property modeling.

Due to the large model uncertainty that still exists with most material models, it is of considerable interest to be able to perform accelerated testing for the identification of proper models and estimation of their parameters. Sensitivity results from preliminary analysis as well as the principles of experimental design theory should be applied.

In general, wood structures are periodically inspected and their maintenance is performed. The maintenance cost is a significant part of the total life time cost. For some steel structures, inspection and maintenance planning is derived from stochastic analyses and substantial savings have been achieved without compromising the safety and serviceability. This optimization has been based on general stochastic models which should be considered for wood structures also. Cyclic behavior of wood members, systems, and connections is also a significant problem for wind and earthquake resistant design.

A3. SPATIAL-DEPENDENT BEHAVIOR

In contrast to most other structural materials, timber exhibits considerable variation in strength and stiffness within the structural element itself. This variation has important consequences from a reliability point of view. For instance, the strength (as it is determined by elementary strength of materials formulas) depends on the length of the loaded element and the load configuration. Thus, in a strict sense, the characteristic strength only applies to a population of specimens with the same size and with the same loading configuration as that used in the tests. Such effects are not negligible, as doubling the length of a beam may reduce its moment capacity by 20 to 30 percent. Similarly, the capacity for a beam loaded with a point load at mid-span can be 30 percent higher than if the same beam is loaded with a constant moment along its length. Even more pronounced effects can be found in practice, e.g. in trusses.

To achieve a rational design practice with uniform reliability levels, these effects have to be considered. Various statistical methods are available to study the effects of within-member variability:

a) Discrete random field methods (the timber element may, for instance, be assumed to contain weak zones with random location and random strength.)

b) Continuous random field methods.

Statistical data needed for such studies include:

- Location and frequency of occurrence of low strength zones.
- Distribution and degrees of correlation of strength with spatial parameters.

Locations of low strength zones can be deduced from existing information about stiffness variation, e.g. from grading machines. As an alternative or as a complement, visual observations of defects may be utilized. But data on the distribution and correlation of strength within elements is lacking. Experimental investigation designed to obtain such data should have a high priority.

Once the necessary tools are made available, reliability studies of timber elements in various typical practical situations should be conducted. Further, models accounting for within-member variability should be combined with analysis of duration of load and system effects.

Research based on the same philosophy should be made for connections, as well as for engineered products such as glued laminated timber and laminated veneer lumber.

A4. RELIABILITY BASES FOR DESIGN

A4.1 Importance

Measures of structural reliability, such as the reliability index, β, or limit state probability, p_F, provide quantitative measures of safety and serviceability if only in a relative and rational sense. Accordingly, the concept of a target reliability measure is fundamental to the development and implementation of probability-based limit states design procedures, regardless of the level selected. At Level I, the partial factors of safety on loads and material strengths are selected to correspond (approximately) to the desired level of reliability; at Level III, it is incumbent on the design to demonstrate that the reliability associated with a proposed structural solution is at or above the desired value. These reliability benchmarks must be made available to the code-writing groups or to practicing engineers who may wish to use them as a basis for designing unusual or high-hazard structures not falling within the scope of the code.

A4.2 Current Status

Because of the evolutionary nature of design codes, measures of reliability associated with existing acceptable design requirements are felt to provide benchmarks for probability-based limit states designs. Accordingly, considerable effort has been directed toward establishing these benchmarks through evaluation of existing member design criteria. This has been done for steel, reinforced concrete, cold-formed steel, and now most recently for engineered wood construction. The emphasis has been on member design criteria for flexural, tension, and compression limit states, which are reasonably well understood. The first generation of probability-based limit states design requirements are based on these calibrations. The probabilistic models of loads and strengths are based on limited data, leading to a statistical modeling uncertainty, which is not considered explicitly.

A4.3 Limitations of Current Knowledge

Certain limit states have received little attention in the calibration exercise because they are not as well understood as those mentioned above; these would include limit states for many types of connections. Moreover, system effects, if they exist, have not been considered in most calibrations. Because of this incompleteness, sensitivity problems in structural reliability, and the role of human error, it is difficult to relate the reliabilities calculated to the observed failure rates.

While there is evidence of a slow convergence in time toward socioeconomic reliability levels, one cannot assume that this has been obtained for individual construction materials. The question of whether optimum reliability levels for a specific application have been attained has not been answered in current code calibrations. Moreover, mechanisms for incorporating feedback from in-service performance have not been included in a formal way. Thus, the inclusion of such information tends to be anecdotal in nature.

A4.4 Needs

A more rational method for setting target reliability measures is required for both ultimate and serviceability limit states. A minimum cost optimization procedure would be desirable in the long term. However, the information needed to do this presently does not exist for all limit states. Specific observations of the Working Group include:

- Serviceability limit states--minimum life time cost model could be implemented

- Ultimate limit states--the need to include cost of human life/injury is a barrier to the use of minimum cost optimization

Objective functions for minimum cost analyses need to be designed. These functions will depend on the conflicting viewpoints of the participants in the building process; therefore, a multi-objective optimization may be necessary.

Existing practice is the best natural laboratory to view the consequence of code decisions. However, a systematic way of feeding back this experience into the code development process does not exist today. Furthermore, when new products are introduced, a new service environment is encountered, or when new knowledge is obtained, techniques must be developed to accommodate revisions accordingly. Mechanisms must be developed to filter this information and to revise reliability targets. This process seems more straightforward for serviceability than for ultimate limit states.

Many building products are designed by testing rather than by calculations. This is particularly true in the wood industry, where roof trusses and similar systems may be "designed" this way. There is no way at present to ensure that the reliabilities of structures and components designed by these two methods are equivalent. To the end user, these reliabilities should be the same. New methods must be developed to ensure consistent reliability regardless of which method is selected. A protocol for design by test would include a description of component boundary conditions, standard loading conditions, the spectrum of test conditions, and proper physical interpretation of results.

Statistical data required as the basis of reliability analysis is subject to a wide range of interpretation. Data invariably are limited in reliability analysis, and different analysts may obtain different results from the same data. A standard protocol is needed to assume that different analysts obtain the same results. This is especially important for level III design, where full distribution methods are used and specific reliabilities are calculated. One way to accomplish this is through the use of standard distributions for specific load and resistance variables. However, the use of such distributions may place undue constraints on analysts with a high quality data set which may not conform to the standard assumptions.

A5. SERVICEABILITY LIMIT STATES

A5.1 Importance

Serviceability limit states are becoming increasingly important in modern construction. The fact that wood has a relatively low stiffness-to-strength ratio results in relatively flexible structural systems. Serviceability limit states for wood construction involve excessive short and long-term deflections that may cause unsightly sagging and damage to finishes and nonstructural attachments, as well as objectionable vibrations due primarily to the activities of building occupants and connection slips. The focus in code development in most countries has been on the ultimate or safety-related limit states because of the traditional concern of code groups on public safety. As a result, serviceability requirements, when they exist, may be insufficient to ensure acceptable building performance in service.

A5.2 Current Status

There are few serviceability guidelines in current codes. Most existing provisions are in the form of a limit on static deflection as a fraction of span (l/360 under full nominal live load is common) or limitations on span-to-depth ratio of flexural members ($l/d<20$ would be typical). Such requirements are aimed at controlling stiffness and, in fact, limit the curvature of flexural members.

A5.3 Limitations of Current Knowledge

The current requirements work reasonably well when serviceability is primarily related to stiffness of the structure. For example, damage to finishes and nonstructural attachments often occurs at deflections on the order of 1/200 - 1/300 times span or story height. However, the requirement to perform deflection checks using the full nominal load may be excessively conservative in some instances, particularly when the condition of in-serviceability occurs for only a short time and is nonpermanent (reversible) in nature. Current serviceability requirements do not differentiate between those conditions that are reversible and those that are not. Moreover, limitations on structural stiffness are insufficient to control objectionable structural vibrations.

A5.4 Needs

(Not in primary order).

- Quantitative deflection limits associated with failure of floor finishes, nonstructural components, and moisture penetration of the building envelope.

- Quantitative deflection limits associated with visually objectionable deflections.

- Structural motion limits (deflections, accelerations) associated with objectionable vibrations.

- Stochastic load models required to analyze long and short-term serviceability.

- Cost models to be used in optimizing life-cycle costs of serviceability.

- Load combinations for checking serviceability, derived from prescribed probability of exceedance (occurrence).

- Minimum life-cycle cost.

- Code implementation mechanism.

A6. BEHAVIORAL MODELS

From the simple calculation of an elastic single element up to reliability calculations of complex systems, different improvements might be useful:

- More representative constitutive equations of the materials, i.e. taking into account different non-linear (elasto-plasticity, visco-elasticity, large deformations....) behavior.

- Investigation of different modes of failure which is equivalent to the extension of limit state equations (damage approach, fracture mechanics under combined stresses,...)

- Generate a better knowledge on the interactions between members, i.e. formulate local equations for modeling the connections.

From the material point of view, researchers have investigated more and more sophisticated constitutive equations. But this information is usually very difficult to incorporate into a structural analysis program due, for example, to computer limitations. Therefore, reliability analysis should be used as a tool to evaluate the importance of model uncertainties through sensitivity analyses and to give more guidelines for incorporating progressively sophisticated models.

An example of the interaction between material behavior and structural analysis is clearly the problem of connections. From the simple hinge model to a stochastic finite element model, different stages of complication might be reached. The guideline for this procedure might be to analyze the influence of the assumptions of this model as a function of the global behavior of given structures. This analysis should not only compare predicted and observed data, but also look at the sensitivity factors related to the model.

One consequence of the above approach could be to obtain a better idea of the ductility of both the timber and its connections and therefore to take benefit of the extra safety which exists in structural systems. Another advantage is to provide a feedback to material science for orientation of the models which are most critical in the behavior of structures.

A7. TOOLS FOR RELIABILITY ANALYSIS

We believe that where fundamentals of reliability assessment are concerned, the need is not so much the development of new tools, as it is the dissemination of the current state-of-the-art to end users. In particular, the following tools are important.

a) Sensitivity measures of the design process can be used to assess reliability where the need for information is the greatest. This can direct research funding into areas of greatest need by putting proposals into the context of the total design process rather than just letting research proposals follow the availability of funds. Sensitivity factors, such as the relative magnitudes of the safety margin coefficients, the omission of sensitivity factors, and derivatives of the reliability index with respect to distribution parameters, can be cited as evidence that the proposed research will help to bring our information levels into balance without over emphasizing any single variable.

b) Some problems cannot be dealt with by simple random variable models. Such problems as size effect, load configuration effect, and load/slip analysis of multi-tooth connections cannot be treated with random fields. However, spatial variability of members and connections within an assemblage, or variations of properties within a single member (which is characteristic of wood) can be treated by the stochastic finite element method.

c) Time-dependent reliability methods are available to treat such problems as:

- first crossing of load variable
- damage accumulation (first crossing of damage state variable)
- residual strength analysis

The specific research need in this area is for a validated mechanistic model which represents the dependence of strength upon time. It should take into account the mechano-sorptive effects of moisture movement upon creep rate. It would be very valuable to pursue the time-temperature dependence of wood, as predicted by kinetics of polymer chemistry, to enable accelerated tests for creep and creep-rupture (below some suitable maximum temperature).

d) Bayesian reliability analysis methods are attractive to civil engineers because of their freedom from frequentist interpretation. They could be used to periodically update resistance distribution, using the existing prior distribution, and adding new data which reflected changes in the forest resource. An appropriate standard, or addendum to an existing standard, would be needed.

e) Finally, it should be noted the lack of available computer software is no longer the impediment it once was to the fruitful application of these tools. Many fine software packages are available and more may be expected.

A8. EDUCATION AND COMMUNICATION

One of the main objectives of developing RBD codes is to correct the inconsistent assignment of loads and resistances, define the variation in the actual reliability of various structures, and to provide for a known, more accurate and consistent reliability across different structures.

To this end in goal, designers, structural researchers, and code writers have to communicate with each other. Practicing engineers must be able to view a design problem with a stochastic approach and relate past experience and judgement to reliability-based targets. That is why proper knowledge of structural engineers in probabilistics and statistics is needed. To attain that, courses of theory of probability appears necessary in the curricula of structural engineers. These courses, beyond what may be generally taught today, should include topics of probability useful for representing and evaluating test data of wood and connection properties, the stochastic nature of load, and the environmental effects. These studies should also be aimed at giving theoretical basis and skill in reliability analysis, as well as decision analysis in general. Another aspect of the communication needed between structural analysts and code writers is the unique behavior wood structures, as compared to structures made of other engineering materials.

Understanding of the behavior and proper models of wood and its connectors may help code editors, designers, and researchers to work together more efficiently. It emphasizes the need of teaching these in-depth wood-related issues to structural engineers and to make this information available in a consistent and systematic form for code writers.

A9. CONCLUSION

For the assigned topic, the Working Group identified several major recommendations for further research:

- Design accelerated life tests in interaction with structural reliability analysis.

- Take into account the environmental influences (temperature and moisture as load effects) on the duration of load behavior on dimension lumber.

- Develop knowledge on cyclic behavior of members, connections, and systems.

- Collect information of within member strength variability.

- Develop models for multiple fastener connections, in relation to structural behavior.

- Integrate life cycle cost models for evaluating target reliability, especially for serviceability.

- Develop methods to compare reliability levels of structures designed by tests and by analysis.

- Set up a protocol to insure a consistent treatment of uncertainty information (statistical models).

- Integrate maintenance policies with reliability-based design.

- Investigate load combination based on predictability of exceedance and minimum life cycle costs.

- Investigate the residual strength models based on past load/moisture history.

- Extend the use of sensitivity analysis for evaluating the uncertainties of behavioral models.

Because most of the fundamental tools of reliability analysis are available, the improvement of RBD of wood structures requires a better use of those tools. Therefore, it is necessary to implement these methods into education programs of civil engineers, extending the current scopes of statistics and predictability units.

Another issue for a better knowledge of RBD is to improve communication. A working group of CIB W18 on reliability would give an opportunity for different countries to share concepts specific for timber structures. It is also essential to establish closer interactions among different materials. As an example, the integration of timber engineering in the Joint Committee on Structural Safety would help to enhance the common use of fundamental theories of reliability-based design.

GROUP B: RELIABILITY ASSESSMENT OF MULTI-MEMBER STRUCTURES

Organizing Committee Member	:	Peter Glos (unable to attend)	Germany
Speaker	:	Rudiger Rackwitz	Germany
Group Leader	:	Marita Kersken-Bradley	Germany
Rapporteur	:	Joe Murphy	U.S.A.
Participants	:	Marvin Criswell	U.S.A.
	:	Lars Ostlund	Sweden
	:	David Rosowsky	U.S.A.
	:	Erol Varoglu	Canada

B1. IMPORTANCE

Typically, wood structures include many individual components and members acting together at several levels as an often complicated structural system. Although practical design is most usually based on the design of individual members, members and components often act beneficially together (for example, sheathing interacting with joists, rafters, and studs). Overall structural performance can often be limited by the performance of any one or several of the critical elements.

Reliability-based assessment and design of wood systems can provide more uniform safety and may provide more efficient and economic design over single-member analysis. For instance, present U.S. design practice allows single-member bending stresses to be increased by 15 percent for specific parallel redundant systems; the 1990 Canadian limit states code allows an increase of up to 40 percent. The benefits of these types of system actions are not fully recognized in most current codes.

Designers of large timber structures have long included at least an intuitively-based consideration for overall system safety and the identification of important key components in maintaining structural integrity. More explicit design attention to these system safety aspects and code provisions, based on more complete and accurate system reliability assessments, can be key in improving the already generally excellent level of wood structural performance. System analysis can help to design future large, innovative, and/or complex systems using new wood-based components and products.

B2. CURRENT PRACTICE

Apart from the system factors of 1.15 in the U.S. and the system factors of up to 1.40 in the Canadian code, the partial coefficient, γ, approach is adopted by the EUROCODES.

J. Bodig (ed.), Reliability-Based Design of Engineered Wood Structures, 159–167.

$$F_M = F_m \cdot F_{Rd} \qquad [B1]$$

and

$$F_F = F_f \cdot F_{Sd} \qquad [B2]$$

where:

F_M= partial coefficient for resistance

F_m= partial coefficient for material

F_{Rd}= model uncertainty in the resistance

F_F= partial coefficient for load effect

F_f= partial coefficient for load

F_{Sd}= model uncertainty in the load effect

Once these partial coefficients are computed, then they can be used to modify material resistance and load effects. In particular, F_f, is used in a linear effect format to compute the design load effect, S_d, as

$$S_d = S\,(F_F\ F_k) \qquad [B3]$$

or in a non-linear effect format

$$S_d = F_{Sd} \cdot S\,(F_f\ F_k) \qquad [B4]$$

where

F_k= characteristic or representative force

Mechanical system effects are accounted for in the F_{Rd} and F_{Sd} partial coefficients. In principle, system reliability aspects could also be accounted for in these factors.

In Australia, the modification factor, k_{sys}, for system effects can contain two distinct components. One is the factor arising from the material characteristics, particularly its variability. This is the only effect considered for k_{sys} in AS1720.1. The second component (that might be included in k_{sys}) is due to idealization used in structural analysis. In AS 1720.1, the structural analysis is assumed 'exact'.

B3. GENERAL DEFINITIONS

To properly discuss system reliability, some basic notions need to be defined. A system is understood as an arrangement of "elements" which interact mechanically. In turn, the elements are associated with topological or functional elements or modes in which an element can transit from one state to another. So far, elemental and system states are categorized into only "safe" and "failure" states, separated by the limit state function. Thus, a structural element is associated with a single state function. System elements do not necessarily correspond to structural members.

Systems may be classified into redundant and non-redundant systems. In general, it is sufficient to identify a non-redundant systems by a series system. Redundant systems may in turn fail in several modes which form a series system of system failure modes. Therefore, the general set theoretical representation of a system can be in terms of a (minimal) cut set of elemental state change events.

System analysis using the same mathematical concepts can be performed not only for the overall system but also on the level of sub-assemblies, connections, etc. and for the elements themselves when an appropriate redefinition of the "elements" is enacted.

B4. SYSTEM BEHAVIOR

System behavior may be visualized, for example, according to Fig. B1, where the passage of a structure through different states is shown; the serviceability state, a state of local failure, and a state of system failure (collapse). Case I represents system failure without a previous local failure, e.g. a failure due to instability of the whole system. In this case, the whole system is, in principle, functioning as a single component. Case II identifies component failure and system failure which occurs in the same process. The system is a typical series system (weakest link system). In case III, a local failure occurs first which causes a change in the structural system. The system failure concerns this modified system and in principle the local failure and the system failure are two different kinds of failure separated in time. There may also be several successive local failures before the final system collapse. Case III is typical for progressive collapse. In case IV, the process stops after the local failure, no system failure occurs, and the structure is sufficiently robust.

The potential failure types I and II can, generally, not be avoided in the design. The structure should be designed so that it has sufficient degree of reliability against these kinds of failure. The failure type III should be avoided as far as possible. Case IV represents the most desirable arrangement.

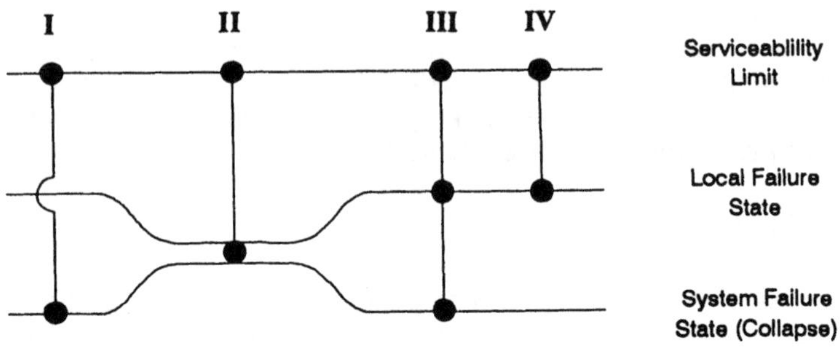

Figure B1. Conceptual Visualization of System Behaviors

B5. PRACTICAL DISTINCTION

For design purposes, systems may be subdivided into sub-systems whose failures are defined as system failures.

B5.1 Global Structural Systems

Global structures systems include sub-assemblies and may include the whole wood structure. This global system would include consideration of interactions and performance of sub-assemblies and the general performance, for instance, progressive collapse of the whole structure.

B5.2 Sub-assembly Systems

Sub-assembly systems including but not limited to:

- Components (e.g. prefabricated trusses)
- Fasteners
- Shear wall and diaphragms
- Floor systems
- Roof systems
- Wall systems

It is particularly important that we know the structural properties of sub-assemblies (load-bearing capacities and stiffness). Those properties should be assessed by system analyses conducted by researchers and transferred to simple code requirements.

B5.3 Micro Systems

Micro systems are systems (for determining size effect) which reflect material variability within a member's volume. An example would be a knot located not at maximum stress in a member. We will not discuss this system further since we believe it should be in the scope of Material Characterization Working Group.

Formally, the same system reliability methodology can be applied for serviceability limit states.

B6. LIMITATIONS OF CURRENT CODES

In principle, structural codes are single-member design codes. System aspects are not taken into account. Generally, codes do not distinguish system reliability levels which depend on the consequences of component failure.

B6.1 Global Structural Systems

Presently there are few guidelines or criteria for assessing the performance of global systems or identifying structures that might be susceptible to progressive collapse or disproportionate damage. The U.K. code and U.S.A. load code contain provisions for progressive collapse.

B6.2 Sub-Assembly Systems

The system factors in the U.S.A. and Canadian codes do not explicitly identify whether they include analysis for composite action, two-way action and/or statistical consideration (i.e. variability between elements). Composite action can be accounted for deterministically on pure mechanics principles. Also, those systems factors only apply to a limited few sub-assemblies. Provisions are missing for any other systems.

B7. OBJECTIVES OF SYSTEM CONSIDERATION

The target reliability of a system should be related to the effort to achieve reliability and to the consequences of system failure. This, in turn, determines the target reliability for the elements for subsystems.

Target reliabilities should be related to the structural system and not to individual components independent of their function within the system. Depending on the type of system, different reliability levels may be appropriate for individual components and for different component failure consequences. For example, non-redundant systems may require higher component reliability levels. These higher reliability levels may be achieved by design provisions and/or quality assurance measures.

Guidelines should be developed for the classification of structures as, e.g. redundant and non-redundant, and for the assignment of components to safety classes.

Structures should be designed such that they are not excessively damaged due to influences not explicitly covered in the design. This implies that structures are robust or insensitive to:

- Loads, including spatial variation.
- Imposed or constrained deformations (e.g. differential settlement).
- Chemical/biological influences.
- Idealizations in modeling (component, system).

Furthermore, structures should be designed to minimize the risk or the consequences of progressive collapse and human error.

Satisfying the different criteria for robustness may result in conflicting design requirements and it is the responsibility of the designer to select the appropriate strategies.

B8. RESEARCH NEEDS

B8.1 Global Structural Systems

The safety of the entire structure or a significant portion of a structure depends greatly on the choice of structural configuration, member as well as element, strengths and ductilities. Structural detailing, including bracing and connection details, also affects system reliability. The sensitivity of a structural system to an individual component failure due to inadequate alternate load paths must consider:

- the structural layout
- the type of structural system
- the resulting degree of redundancy
- the stochastic dependencies of element properties
- the ductility of both members and connection regions

A key research need is to provide methods for identifying when the level of structural system robustness and redundancy is high enough that explicit designer attention to progressive collapse concerns is not needed. Depending on the architectural and functional requirements, it may or may not be economical to choose such a system.

Obtaining effective moment connections is generally more difficult in timber structures than with many other structural materials. Research on practical ways to achieve more redundancy and ductility on engineered timber systems is needed.

If a fail-safe (i.e. very low possibility of progressive collapse failure mode) is chosen, research is needed to provide information on the details (bracing, connections, etc) required, so that such a robust system is indeed achieved.

If a system with significant series behavior is chosen, adequate analysis and design procedures need to be better defined for both the researcher providing input to codes and, eventually, for designers. A procedure is needed for identifying the components and members which are the most critical in controlling the system reliability. Research is needed on how to accurately identify these critical members, including the number of such critical members for a given system, as a minimum number may be desired to minimize the effort for later design steps.

Procedures for identifying critical members can consider that many components and members are often considerably oversized due to the limited number of standard sizes or product types. Oversizing may result from the designer decision to provide uniformity of sizes for similar members and from other pragmatic concerns. Structures designed so that many or most of their members are close to their individual design capacity may require particular research attention, as their level of useful reserve strength may be very low even if the structure is statistically indeterminate (Case III in Figure B1).

Possible design approaches for sensitive structures (in the sense of progressive collapse concerns) need to be fully described by research and may include at least:

A. Consideration for the overall system as a series system, in which the more central members and components are identified, and providing higher individual reliability through more demanding member design provisions and/or increased quality control inspection and other field requirements.

B. Sequential analysis of the structure with each critical element, one at a time, is transferred into a failure state and the strength of this modified system determined. This modified system could be required to have a reduced overall reliability and be analyzed with different load factors and/or load combinations than used for a fully effective structure. Increases in member size would follow if second member and subsequent failures are indicated.

Research is needed to define the required increased individual member reliability level (Approach A) and the modified loading requirement (Approach B). These provisions should consider:

- the size and importance (including occupancy) of the structure
- the amount of the structure involved in a potential general failure
- the expected warning of failure and time progression of failure
- other conditions

In all cases, research is needed to better define and then communicate to designers, information on details which can be used to increase the ductility and robustness of series-type structures so that their general performance can be best increased at least cost. This information on details (bracing, connection, added members, methods to provide warning) might first be contained in designer guidelines with some possible codification of at least principles to follow.

Research is also needed on basic structural reliability analysis techniques, including further development of failure analysis approaches which properly model the characteristics of engineered timber structures, their load paths, and load histories. In particular, formulations and solution methods for systems under time-variant, possible dynamic loading, and with load-dependent and/or load independent degradation of system properties need to be studied. One promising research area is the replacement of the Boolean state (zero-one) definition with a continuous state representation. These basic analytical tools based on system reliability theory are needed to provide the best possible information from which the more designer-oriented procedures may be formulated and the adequacy of these procedures later studied for more application-oriented research efforts.

Another overall need across the entire area of structural reliability assessment is a more general agreement on the definitions of such key terms and concepts as robustness, redundancy, fail-safe, and progressive collapse (or, conversely, structural integrity) for use by both the research and the design community.

B8.2 Research Needs

For specified sub-assemblies, reliability analyses are needed to determine structural properties. The analyses must include:

Mechanical Model

- Mechanically correct and representative model.
- Model must converge in the mathematical sense.
- Model must include all relevant failure modes.
- Correct material/fastener behavior (e.g. duration of load, stress-strain relationship).
- Dynamic redistribution of load after local failure (inertia effects).
- Account for representative load paths.

Stochastic Model

- Variability and correlation of random variables (e.g. lot vs. global, between to within elements).
- Spatial variability of loadings which may affect limit states.

- Must include uncertainties in loads, resistances, stiffnesses, geometries, and any due to the mathematical relationship used.

Design Assumptions

Sub-assemblies subject to optimization in a deterministic sense shall be analyzed as being optimized (e.g. trusses).

This level of sophistication in mechanical and stochastic modeling provides for the possibility of counter-acting effects of the assumptions.

The researcher is required to develop simplified design rules for codes (e.g., derive system factors for the above specified sub-assembly.) In the case where the system effects is deleterious, the researcher should investigate possibilities/conditions for which the deleterious effect is mitigated or reduced.

B9. FUTURE ACTIONS

Prior to codification of system reliability results, research is required. Advances are required both in mechanical behavior models and system reliability methods. Research is to be carried out by both timber mechanics experts and reliability experts.

It is recommended that NATO support/facilitate cooperating international research in these areas.

GROUP C: MATERIAL RESISTANCE CONSIDERATIONS

Organizing Committee Member	:	Dave Barrett	Canada
Speaker	:	Ricardo Foschi	Canada
Group Leader	:	Bob Tichy	U.S.A.
Rapporteur	:	Roger Lovegrove	U.K.
Participants	:	Jean-Pierre Biger	France
	:	Ario Ceccotti	Italy
	:	Vijaya Gopu	U.S.A.
	:	Cathy Marx	U.S.A.
	:	Takashi Nakai	Japan
	:	Lauge Nielsen	Denmark
	:	Juhani Taipale	Finland

C1. INTRODUCTION

The rational design of wood structures requires material resistance assignments based on at least some knowledge of material properties.

Traditionally, material resistances have been defined using data largely derived from small, clear specimens and then modified to take into account the influence of a variety of factors such as natural defects. Recent research has demonstrated that with full-size test data, improvements can be made in the development of behavior models* applicable to a wide range of species, grades, sizes, and types of products. Thus, the development of information about materials properties appropriate to full-size members remains one of the major means of improving our structural behavior models.

Timber structural design codes must evolve from the allowable stress design (ASD) to the limit state design (LSD) code formats if they are to remain comparable to, and as acceptable as, the other major structural materials which are already being designed in accordance with LSD.

It is generally agreed that transformation from ASD to LSD will be aided by access to material property data and structural models based on full-size component and system studies. These studies must establish the in-situ behavior of timber products for a wide range of in-service environmental conditions.

New structural wood products will allow timber to compete more effectively with other structural materials. These new products tend to posses lower variability and higher design properties, as well as greater uniformity, than traditional structural wood products. Typically, they are subjected to high levels of quality control.

---

*By "models" is meant the representation of both test data and behavior in a mathematical form suitable for engineering analysis of different limit states.

J. Bodig (ed.), Reliability-Based Design of Engineered Wood Structures, 169–175.

The evolution of structural wood products requires that they be equitably rated in relation to all competing materials. Reliability-based design (RBD) provides a framework for those safety studies necessary for rational design in wood, assuming the required materials-properties database is available.

Understanding the behavior of connections is crucial to establishing the behavior of wood structures. Analysis of connector behavior is hampered by a lack of knowledge of the wood failure modes, particularly under combined stresses.

Structural wood products are extensively traded on the international markets. Opportunities for expanding international trade will increase as international trade barriers are removed. These trends suggest that even greater emphasis should be placed on international cooperation in materials-properties research programs and in the development of design codes.

C2. STATE-OF-THE-ART

The world-wide trend in wood engineering has been towards the development of materials-properties from full-size testing together with the development of product-independent performance-based standards for product evaluation.

In general, there is now a good understanding of the static performance of individual members under controlled conditions, as well as the existence of a basic knowledge on the microstructure of wood.

Overall, ample data exists compared to other construction materials such as concrete and steel. For full-size lumber, extensive static testing has resulted in excellent bending, tension, compression parallel-to-grain and modulus of elasticity data. However, only limited information has been collected for shear, tension, and compression perpendicular-to-grain. Creep, fracture and environmental effects, such as moisture, temperature, and fire have received reasonable attention while information about the behavior of connections and adhesives remains minimal.

Other structural wood products, such as glued-laminated timber, panels, I-beams and laminated veneer lumber, have been investigated to varying degrees.

There now exists significant experience in the design and construction of wood products in most end-use markets. This includes new construction and the repair and remodel markets, for both residential and non-residential buildings, along with other structures, such as wood bridges and concrete formwork.

Application of statistical evaluation techniques is not new to wood structures. Today the main challenge lies in their uniform application. Various countries have developed reliability-based procedures for wood structures, independently of one another, resulting in inconsistencies in the procedures for developing material resistances. Techniques for translating from one system to another are now needed for immediate use.

C3. RESEARCH NEEDS

The quest for knowledge of materials-properties must be driven by the need or structural modeling in a reliability context. Data are needed for different species and for different environmental regimes.

In this context, the following needs can be identified from among others:

- Quantified size effects in strength and stiffness. These effects need to be studied in individual members in tension, bending, and compression. This evaluation needs to be linked with within-piece variation of properties.

- Nondestructive evaluation, and its procedures, must be further improved with a view to linking them with reliability-based procedures for more efficient grading or classification of timber. Here, the needs of the end-user should always be kept in mind.

  Additional techniques are required for the evaluation of existing buildings and to aid the correct choice of repair techniques.

- Quantification of shear strength and of tensile strength perpendicular-to-grain as influenced by the size of the member and by the stress distribution. These effects, and their interaction, are also required for the study of wood-failure mode in connections.

- Improvements must be sought in our knowledge of the long-term behavior of non-lumber materials. This information could help improve structural modeling, and also clarify questions of the linkages between strength degradation and creep, as well as the interaction between environmental effects, threshold levels for strength, upper limits of creep deformation, etc.

- The reliability of wood under conditions while exposed to fire in non-residential applications needs further investigation.

- More information is needed about the conditions for both short and long-term crack propagation. This is essential for the analysis of existing structures in addition to being needed in the design process.

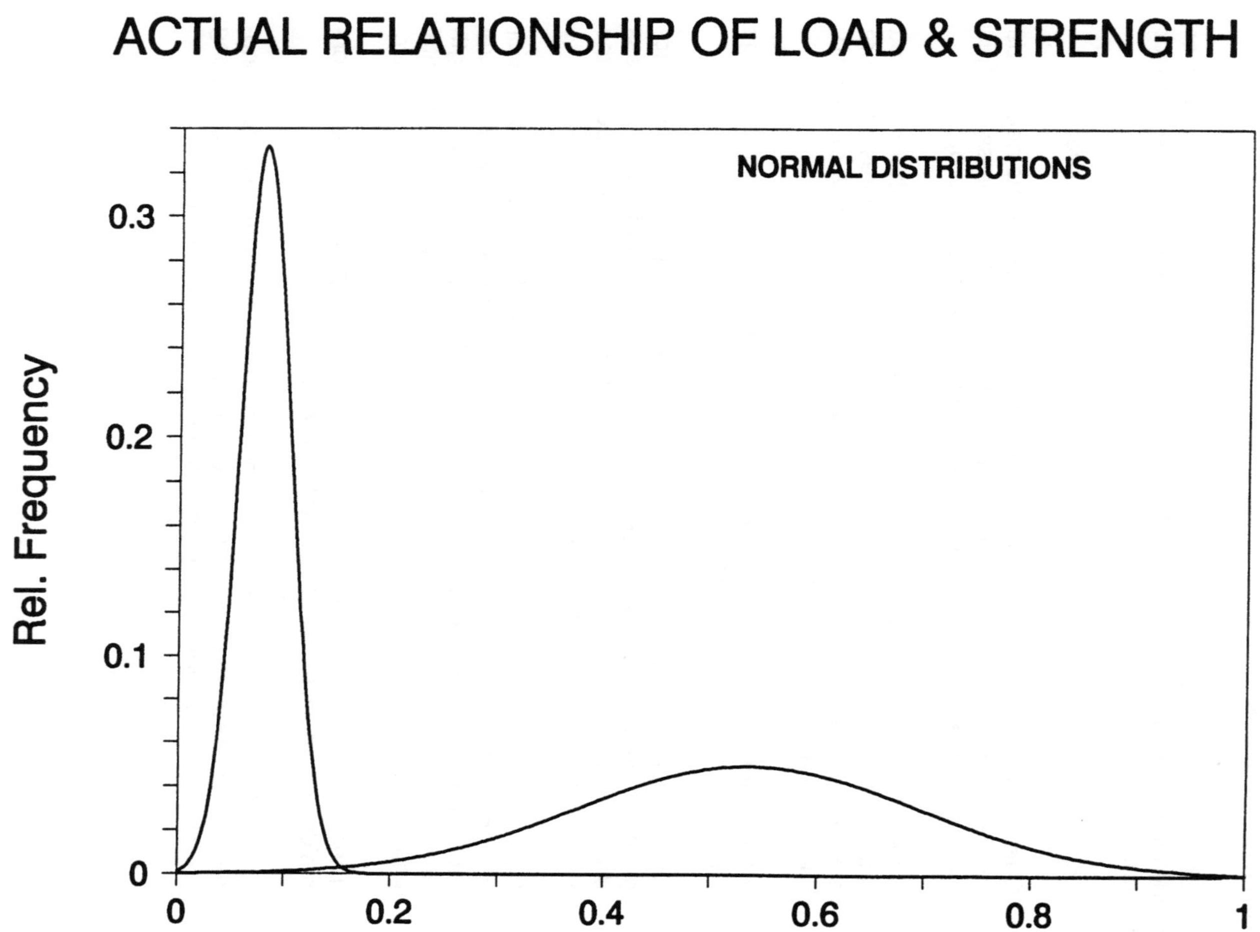

Figure 1. Typical Overlap of Load Effect (Left Curve) and Resistance (Right Curve) of Engineered Wood Structures

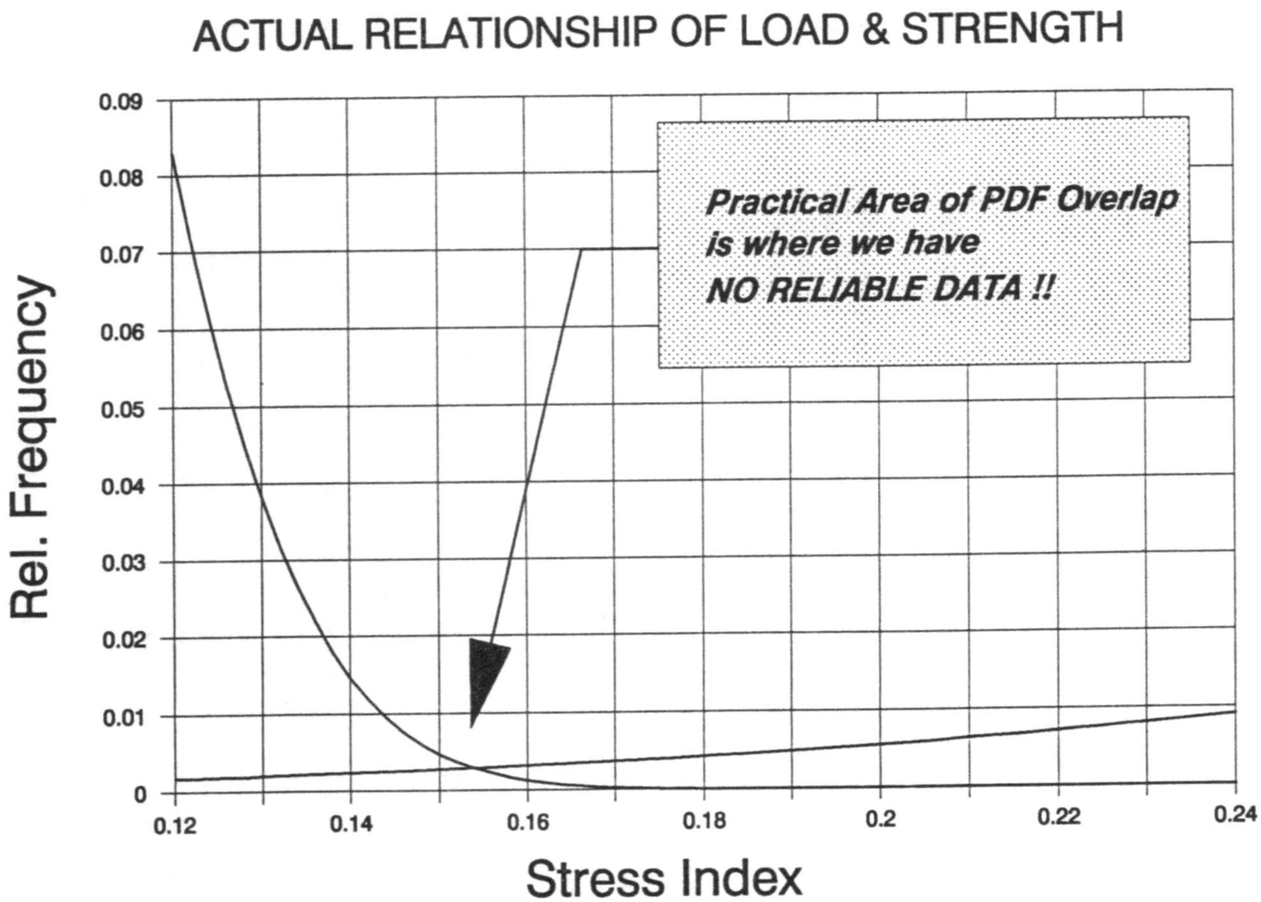

Figure 2. Amplified Area of Overlap of Figure 1.

- Strategies need to be developed for adapting, in an efficient manner, to changes in the quality of the raw material resource.

  Understanding the behavior of connections is crucial to establishing the behavior of wood structures. Analysis of connector behavior is hampered by a lack of knowledge of the wood failure modes, particularly under combined stresses.

- Ductile and cyclic behavior of mechanical connections, resulting from the behavior of wood in compression perpendicular-to-grain and ductile yielding of fasteners, must be studied for dynamic analysis (for example, wind and seismic excitations).

- Appropriate descriptions of materials properties will lead to the development of higher-level analysis models for structural products, such as glued-laminated members, trusses, and I-beams. These models would in turn permit the reliability evaluation of those products using a system approach for the different modes of failure.

Since most practical structural design deals with low probabilities of failure when collecting material resistance data, one should be cognizant of the need to accurately define the low-tail end distribution of values. Figure 1 illustrates the typical overlap of areas represented by the load effect distribution (left curve) and the material resistance distribution (right curve). This overlap region is amplified in Figure 2 showing that in a typical case, the intersection point is located about the 98th percentile of load effect and 0.6th percentile of resistance.

C4. RECOMMENDATIONS

Wood is an international material. It is more extensively marketed on the international scale than any other structural material. This aspect, together with diverse international codes and standards, makes it essential that the degree of global cooperation be much greater for wood than for other construction materials.

The collection and collation of information about wood products must, therefore, be a global effort with input from all partners in the timber community. The problem, then, becomes that of international cooperation and communication. To this end, the following strategic recommendations are made:

- World-wide harmonization of fundamental concepts is required.

- Formation of an international data-base of timber properties.

- Establishment of internationally coordinated effort focused on the specific tasks identified in Section C3.

- The improvement of international communication by:

    - production of an international RBD Newsletter with input from all sectors which have an interest in RBD;

    - establishment of a regularly scheduled RBD forum for the high-level exchange of information.

Throughout, the overall objectives should be to maintain or improve wood's competitive place in the construction market, by clearly focusing on the needs of the end-user, while maintaining acceptable standards of safety.

GROUP D: GLOBAL HARMONIZATION OF TIMBER ENGINEERING CODES AND STANDARDS

Organizing Committee Member	:	Bob Leicester	Australia
Speaker	:	Bob Leicester	Australia
Group Leader	:	Geoff Boughton	Australia
Rapporteur	:	Bryan Walford	New Zealand
Participants	:	Erik Aasheim	Norway
	:	Ron Anthony	U.S.A.
	:	Ad J.M. Leijten	Netherlands
	:	Tom McLain	U.S.A.
	:	Raphael Mutuku	Kenya

D1. INTRODUCTION

Timber is an important global commodity that is used in construction in nearly all nations and in many types of structures. For example, in 1987 in the United States, more wood products (40.4 billion dollar value) were used in construction than all other major materials (metal, concrete, gypsum, and tile) combined (37.5 billion dollar value)*. It is a renewable resource that requires low energy in its conversion to structural components, as well as being a resource which provides essential ecological benefits in relation to the use of alternate materials. These attributes ensure its place as an important structural material for the future.

D2. CODE DEVELOPMENT

The development of structural design codes, in general, is proceeding rapidly with the adoption of the reliability-based design (RBD) concept often utilizing the load and resistance factor (LRFD) format. Another form of RBD is the limit states design (LSD) format. In this respect, timber codes are lagging behind those of other materials with the consequence that timber design is being ignored in many schools of engineering focusing on the RBD methodology. The LRFD is a format which allows the application of reliability analysis to improve consistency of safety levels between various types of structure, loadings, and construction materials. Further, LRFD facilitates the introduction of new technology and new products more easily than previous formats. For all these reasons, it is imperative that timber design codes be converted to the LRFD format as soon as possible. Without such conversion, reduction or even total loss in the utilization and trade in structural timber will result.

*1991 United States Industrial Outlook. U.S. Department of Commerce. U.S. Government Printing Office, Washington, D.C.

J. Bodig (ed.), Reliability-Based Design of Engineered Wood Structures, 177–183.

Timber design codes are being developed differently in different regions, depending on the trade patterns and technologies characteristic of those regions. For instance, the European Community (EC) is about to publish a suite of codes which will act not only as a stimulus (or lubricant) to trade in both materials and expertise between member nations, but also in some respects as a barrier to non-member nations. An LSD code has already been published in Canada and others are being drafted in U.S.A. and Australia. Without code harmonization of structural design standards, timber remains a local or regional commodity rather than a global one. Differences in standards for material evaluation leads to duplication of testing and evaluation, thereby increasing costs and producing delays. These differences also inhibit technology transfer between countries. This constraint can be particularly damaging to developing countries that rely on developed countries for their technology. For these and other reasons, the lack of harmony between timber codes and standards acts to diminish the viability of timber as a global commodity and may eventually reduce the world-wide use of this excellent renewable engineering material in favor of man-made materials such as steel, concrete, ceramics, or plastics.

D3. CODE HARMONIZATION

Harmonization of timber standards implies compatibility rather than uniformity. Codes of differing complexity can be tolerated, and are even desirable, so long as there is a harmony of intent and compatibility of format between them. This is particularly relevant to developing countries where timber construction represents very appropriate technology and the sophistication of the code needs to be matched to that of the supporting manufacturing infrastructure.

D4. STATE-OF-THE-ART

The current position is summarized in the sub-topic paper: The Development of LSD Codes for Structural Timber by R.H. Leicester. Currently, a few countries have operational LSD codes for timber engineering and there are several with draft codes that are nearing completion. In terms of basic format, these codes are similar. The formats are deterministic and involve partial safety factors which have often been derived through the use of reliability studies. There are, however, several significant differences between the codes of various countries. The most significant difference involves the methods specified for testing timber in order to measure its characteristic strength and stiffness; these differences are such that it is not possible to predict the relationship between the characteristics values obtained by the measurement procedures of different countries. This not only has a direct impact on the trade of structural timber, but also on the transfer of structural timber technology, such as that related to predicting the effects of load duration, size factors, etc. since these parameters are usually stated in terms of their effects on characteristic values.

Other differences noted were to be found in the methods used for grouping the partial safety factors for resistance; the strength grouping classes used in various regions; the methods used for taking into consideration the consequences of failure; and the effectiveness of quality control procedures.

D5. ADDRESSING THE NEEDS

The incompatibilities apparent in current codes need to be selectively addressed. In order of decreasing priority, it is suggested that the following aims should be targeted:

- To achieve compatibility (but not necessarily uniformity) between codes so as to enable information transfer between the codes.
- To ensure that code formats can be adapted to differing levels of complexity/simplicity in the technology infrastructure.
- To assure that any changes to timber codes are reliability-based, rational and well documented.

The necessity for harmonization of terminology is implicit in that variables, properties, loads, and treatments must be defined unambiguously. This does not imply identical use of the provided definitions, but sufficient compatibility must be obtained so as to enable translation of data from one code to another. Ideally, definitions must be globally applicable so that particular species or materials are not excluded. The most important current code incompatibilities that must be addressed are listed and discussed in the following:

D5.1 CHARACTERISTIC VALUES (CRITICAL PRIORITY)

All design relates to characteristic values and, consequently, they are basic to reliability-based codes. It is essential that it be possible for these values to be translated from one code to another. This requires agreement on sampling, specimen configuration, and method of testing. Adjustments to common dimensions, moisture content, sample size, and method used for processing the test data are less critical, provided they are identified and the raw data from tests are accessible.

D5.1.1 Initial Action

The initial action to meet this need is to reach agreement on test procedures so that equivalences can be established between the various code formats.

A workshop on characteristic values is needed to:

- Identify similarities and differences in procedures.
- To develop translation algorithms.

If translation algorithms can not be developed, then proposals should be given for modifying existing testing procedures so as to make the translation possible.

D5.1.2 Long-term Goal

Obtain agreement on common procedures for the definition and evaluation of characteristic values.

D5.2 STRENGTH GROUPING (URGENT PRIORITY)

All timber strength data can be organized into groups, although this entails compromise where the strength group properties do not exactly match those of particular species. Strength grouping allows the easy inclusion of lesser known species and of new products. It also has the potential to improve trade and technology transfer by relating the groups in the various codes.

D5.2.1 Initial Action

Research must be commissioned to find the equivalences between the various existing strength groups, and to determine common rules for assigning timber products to strength groups.

D5.2.2 Long-term Goal

Develop common global strength groups.

D5.3 CONNECTIONS (URGENT PRIORITY)

Connections represent the weakest knowledge link in engineered wood design in that there is much less research done on connections than on timber members. Not only is research on reliability aspects extremely limited, but even the research on behavior of connection is fragmented with too narrow a focus. This narrow focus, in part, is due to the fact that the number of combinations of connector type, timber thickness, load direction, and connector orientation is extremely large. There is a clear need for enhanced commonality of design rules for connections with, common agreement on, and an understanding of the effects of various parameters on the characteristic load values.

D5.3.1 Initial Action

A workshop is needed to:

- Find commonalities.
- Collate fragmented data.
- Provide direction for future research, particularly with respect to ensuring compatibility of future research studies.

D5.3.2 Long-term Goal

Develop a method for design and testing of connections that is compatible in terms of reliability with the rest of timber design.

D5.4 MATERIAL MODIFICATION FACTORS (URGENT PRIORITY)

In reliability-based codes, all material modification factors, usually denoted as "k factors", must in general be: related to characteristics values; rationally based; clearly defined; and have their base well-documented. At present, the basis of many modification factors is uncertain, while others are drawn from studies on poorly related materials.

D5.4.1 Initial Action

Research is needed that will:

- Assemble the background on existing modification factors.
- Perform analyses on test data to develop translations for existing factors between the various codes.

D5.4.2 Long-term Goals

Develop modification factor definitions that are compatible between various codes.

D5.5 MODIFICATION FACTORS FOR DURATION OF LOAD AND SIZE EFFECTS (IMPORTANT PRIORITY)

(a) **Duration of load**. A test specification is needed to obtain reproducible results and globally compatible factors. The effects of load duration are of major concern in timber design and are unique to timber as a structural material. There is wide debate on the relevance of climatic history, moisture content changes, load levels, member size, effects of the various structural defects, and types of loading. Specific knowledge is needed to translate effects from a standard test environment to local climatic conditions.

(b) **Size effects.** Although there are considerable data worldwide, there is no global agreement on the concepts related to size effect on the strength of timber products. It is recognized that measured size effects are a function not only of material properties, but also of procedures used, such as the method of grading and the definition of characteristic values.

D5.5.1 Initial Action

(a) Establish a draft specification for duration of load testing. RILEM is recommended as the appropriate body to do this work.

(b) Conduct a size effects workshop to provide global direction on definition, analysis, and codification.

D5.5.2 Long-term Goal

Obtain global agreement on duration of load and size effects.

D5.6 OTHER CONSIDERATIONS (UNPRIORTIZED)

(a) Fire resistance is currently prescriptive, rather than performance-based, and because of this, design procedures are not amenable to development through reliability procedures. There should be an increased emphasis on the use of a reliability basis for fire resistance that is comparable to that used for normal structural resistance.

(b) The durability of timber structures is not taken into consideration in design codes with any sort of reliability basis. At present, there is not even a useful predictive model for durability effects. For many types of structures, particularly structures in exposed locations, this state-of-the-art represents a serious deficiency in design, and needs to be remedied.

(c) Serviceability is often the critical design criterion for timber structures and yet codes generally treat serviceability in an advisory matter. A globally compatible philosophy and design method should be developed for serviceability calculations.

(d) Service environments need to be described in terms of a temperature-humidity history for purposes of both testing and design. Analogous to the procedure used for strength grouping, it is proposed that only a few discrete environments should be selected and used.

(e) Terminology varies considerably between codes providing a potential for confusion and design mistakes. Differences are particularly evident in:

- Definition of slenderness
- Stress/capacity format

- Methods for combined stress analysis
- Format used for system effects

D5.6.1 Long-term Goal

The topics of fire, durability and serviceability all need fundamental developments for codification purposes. In doing this, there will be an opportunity to obtain immediate global harmonization and its associated benefits. Steps to obtain this harmony should be taken.

In the more general topics of timber engineering design, steps should be taken to work towards obtaining global agreement on matters related to terminology and design philosophy.

D6. ACTIVE BODIES

Working groups should be established to take responsibility for the identified Initial Action. Existing international organizations, such as IUFRO, CIB, ISO, NATO, and RILEM all need to be part of an overall global effort to make the proposed harmonization possible. National organizations, researchers, and industry have to allocate significant resources and priorities to the many detailed works needed to attain the real goals which will provide major dividends in the long term.

D7. CONCLUSIONS

Harmonization between structural timber standards and codes of various countries will have a profound effect not only on the trade in structural timber and related components, but also in the transfer of technology between countries. In addition, consideration should be given to the fact that the lack of harmonization could have a disastrous consequence on the global economies of timber as a structural material.

In the short-term, it is assessed that a major impact could be obtained through affecting either a global harmonization or developing ways of translating data and technology between various formats in the four following aspects of timber standards:

- Evaluation of characteristic values
- Definition of strength groups
- Improving the reliability basis of connector evaluation and design
- Standardization of material modification factors

In the long term, all timber code bodies should ensure that revisions in codes and standards move towards greater harmonization between codes rather than away from it. This is necessary to preserve timber as a global structrual material.

SUMMARY AND CONCLUSION

The Organizing Committee (Drs. Jozsef Bodig, U.S.A., J. David Barrett, Canada; Peter Glos, Germany; Hans Jorgen Larsen, Denmark; and Robert Leicester, Australia) selected the topic of Reliability-Based Design of Engineered Timber Structures for a NATO ARW in recognition of its extreme importance of this issue for international commerce. Timber products represent an important renewable construction material which is marketed internationally, unlike concrete, which is locally produced and used. Reliable and efficient use of timber products in engineered construction required the harmonization of design codes used in various countries. This ARW has made an important first step toward the eventual harmonization of reliability-based timber design codes.

A total of 36 attendees, representing 16 countries, participated in the ARW deliberation. These participants represented a variety of experts in the field of mathematics, statistics, structural analysis, engineering design, and material science. Categories of employment included university, government, industry, and private consultant affiliation.

The workshop, first of its kind on the Reliability-Based Design (RBD) of Engineered Timber Structures, focused on the comparison of basic approaches associated with the major national and multi-national design codes. The technical issues were divided into four topics:

A: Fundamentals of Reliability Assessment
B: Reliability Assessment of Multi-Member Structures
C: Material Resistance Considerations
D: Formulation of Internationally Harmonized Codes

Four working groups, covering the above four topics, deliberated on the research and development needs to move toward internationally harmonized RBD codes. In addition to a keynote presentation, the state-of-the-art for each of the above topics was summarized by a speaker. Two supplementary presentations enhanced range of topics covered by the speakers. The papers covering all the presentations are part of this proceeding.

It was recognized that sufficient fundamental knowledge and analytical tools exist to conduct reliability assessments. However, because of their unique mechanical properties, modifications in methodologies are needed to allow assessment of the true reliability of timber structures.

The participants concluded that special focus needs to be given to the long-term reliability of timber structures, especially as their reliabilities are affected by time-dependent behavior, cyclic environment, and biological degradation. Both strength limit states and serviceability limit states should be considered.

Timber products are used in a variety of structural configurations. Unified definitions of structural subsystems are needed so proper mathematical model representations could be assigned to them. It was emphasized that in a typical engineered timber structure, components are interacting at several organizational levels. This multi-level interaction makes it difficult to utilize simple reliability assessment methods and, hence, accurate reliability assessment of entire timber structures are not available today.

Material characterization is a critical component of reliability assessment and reliability-based design. A unified approach to the characterization of strength and elastic parameters of wood products is critical for code harmonization. The harmonization has to begin by the coordination of various mechanical test methods by which numerical values are assigned for material properties. Since sampling procedures, specimen dimensions, test conditions, and the test method itself all affect the obtained property values, the development of unified test procedures needs to be harmonized first.

Unfortunately, to date, very few activities have been coordinated at the international level to allow the development of comparable test data. For example, in the EC countries, lumber testing is conducted with the visually defined worst defect placed at the highest stress location. On the other hand, in North America, lumber pieces are tested in random arrangement. Data obtained in the EC countries cannot be utilized in North America and visa versa. Thus, when a product developed in one region is to be utilized for construction in another region, complete retesting and re-evaluation are required. The duplication of efforts greatly increases the cost and unnecessarily complicates the activities of international commerce.

The overall logistics of international code harmonization was the topic of one of the four working groups. Key areas of concerns were identified and their priorities established. Development and acceptance of common nomenclature, glossary, and data collection format were identified as the most critical issues where harmonization needs to focus. These issues need immediate attention as most countries are in the process of developing or modifying RBD codes. If the current opportunity for harmonization is missed, the process will become more difficult at a later date.

It was unanimously agreed by the participants that the ARW represented a very important first step toward the long and extensive effort needed to harmonize RBD timber codes. The participants suggested follow up workshops, seminars, and short courses to begin addressing a number of key issues.

Industry and government organizations need to recognize the importance of international harmonization of RBD timber codes and provide funds for research and development activities. Harmonization is of benefit to both the timber exporting and importing countries.

The ARW participants recommended that NATO assume a leading role and continue its support for the efforts of harmonizing RBD codes for engineered timber structures. NATO should designate this topic as a critical area of concern and provide longer term financial support for further activities. The organizing committee of this ARW would be interested in continuing its involvement in this important issue.

APPENDIX

PROGRAMME, TIMETABLE AND ORGANIZATION

NATO ADVANCED RESEARCH WORKSHOP: Reliability-Based Design of Engineered Wood Structures
June 2-6, 1991, Hotel Demidoff, Florence, Italy

Day 1 **Sunday, June 2, 1991**
Arrive at Peretola Airport, Florence or at the Florence Central Station (which is the terminal from Pisa airport where some of the connecting flights will arrive from Roma and from Milano). Look for the NATO ARW Representative at Peretola or Central Station.
Free shuttle bus service to Hotel Demidoff, located in the suburb of Florence
Hotel check-in
16:00 Workshop registration
20:00 Dinner

Day 2 **Monday, June 3, 1991**
07:00 Breakfast
08:15 Workshop Registration
09:15 Organizing Committee Meeting
09:45 Meeting of Speakers, Group Leaders, Rapporteurs
10:15 Break
10:30 Welcome, Introduction, Objective of ARW, Work Assignment
11:15 Keynote Speaker: Bruce Ellingwood - Reliability-Based Design Concept (joint session)
12:10 Speaker: Henrik Madsen - Fundamentals of Reliability Assessments (joint session)
13:10 Lunch
14:15 Speaker: Rudiger Rackwitz - Reliability Assessment of Multi-Member Structures (joint session)
15:15 Speaker: Ricardo Foschi - Material Characteristics and Reliability-Based Design (joint session)
16:15 Speaker: Lauge Nielsen - Lifetime of Wood as Related to Strength Distribution (joint session)
16:35 Break
17:00 Bus leaves to Ordine delgi Ingeneri (OdI) for Program and Reception.
17:45 Speaker: Jozsef Bodig - Use of Wood Products for Engineered Structures in North America
18:30 Reception for NATO-ARW Participants by OdI
20:00 Bus returns to Hotel Demidoff
20:30 Dinner

Day 3 **Tuesday, June 4, 1991**
07:00 Breakfast
08:15 Speaker: Ario Ceccotti - Safety Format of Eurocode 5 (joint session)
08:35 Speaker: Ad Leijten - Probability-Based Safety Factors in The Netherlands (joint session)
08:50 Speaker: Robert Leicester - The Development of Limit States Design Codes for Structural Timber (joint session)
09:50 Working Group Assignments (joint session)
10:15 Break
10:30 Working Group Discussions (separate sessions)
13:00 Lunch
14:15 Working Group Discussions (separate sessions)
16:15 Break
16:30 Working Groups Report Outline (separate sessions)
17:15 Working Group Draft Report Preparation (separate sessions)
20:00 Dinner

PROGRAMME, TIMETABLE AND ORGANIZATION (Continued)

<u>Day 4</u>	**Wednesday, June 5, 1991**
07:00	Breakfast
08:15	Review and Modification of Working Group Draft Reports (separate sessions)
10:15	Break
10:30	Formulation of Working Group Recommendations (separate sessions)
13:00	Lunch
14:25	Report by Working Group A (joint session)
14:55	Report by Working Group B (joint session)
15:25	Report by Working Group C (joint session)
15:55	Report by Working Group D (joint session)
16:25	Break
16:40	Discussion of Group Recommendations (joint session)
17:30	Summary and Conclusions
17:45	Closing Remarks
20:00	Dinner

<u>Day 5</u>	**Thursday, June 6, 1991**
7:00	Breakfast
	Hotel Checkout
	Shuttle bus to Peretola Airport or Central Train Station

NATO ADVANCED RESEARCH WORKSHOP

"Reliability-Based Design of Engineered Wood Structures"

LIST OF ATTENDEES

A

Erik Aasheim
Norwegian Institute of Wood Technology
P. O. Box 113
Blindern
0314 Oslo 3 NORWAY

Ronald W. Anthony
Engineering Data Management, Inc.
4700 McMurray Ave., Bldg. A
Fort Collins, CO 80525 USA

B

J. David Barrett *(Organizing Committee Member)*
Department Head and Professor
University of British Columbia
Faculty of Forestry #270
2357 Main Mall
Vancouver, B.C. V6T 1Z4 CANADA

Jean-Pierre Biger
Bureau Veritas
Cedex 44
92077 Paris La Defense FRANCE

Jozsef Bodig *(Director and Editor)*
Engineering Data Management, Inc.
4700 McMurray Ave., Bldg. A
Fort Collins, CO 80525 USA

Geoff Boughton
School of Engineering
Curtin University of Technology
GPO Box U1987
Perth 6001 Western Australia AUSTRALIA

C

Mike Caldwell
National Forest Products Association
1250 Connecticut Ave., NW
Washington, D.C. 20036 USA

Ario Ceccotti
Department of Civil Engineering
University of Florence
I-50139 Firenze
Via Di S. Marta 3 ITALY

Marvin Criswell
Dept. of Civil Engineering
Colorado State University
Fort Collins, CO 80523 USA

E

Bruce Ellingwood *(Keynote Speaker)*
Dept. of Civil Engineering
Johns Hopkins University
3400 N. Charles St.
Baltimore, MD 21218-2699 USA

F

Ricardo Foschi *(Author and Lecturer)*
Dept. of Civil Engineering
University of British Columbia
Vancouver, B.C. V6T 1Z1 CANADA

G

Peter Glos *(Organizing Committee Member)*
Institut fur Holzforschung
Universitat Munchen
Winzererstr. 45
D-8000 Munchen 40 GERMANY

Vijaya Gopu
Department of Civil Engineering
Louisiana State University
Baton Rouge, LA 70803 USA

LIST OF ATTENDEES (continued)

K

Marita Kersken-Bradley
Vilshofener Str. 6
8000 Munchen 80 GERMANY

Zsolt Kovacs
University of Forestry and Wood Science
H-9401 Sopron, Bajcsy-z. u.4
HUNGARY

L

Hans Jorgen Larsen *(Author)*
Danish Building Research Institute
SBI, Postboks 119
DK-2970 Horsholm DENMARK

Robert Leicester *(Author and Lecturer)*
Division of Building, Construction & Engr.
CSIRO
P. O. Box 56342
Highett, Victoria 3190 AUSTRALIA

Ad J. M. Leijten *(Author and Lecturer)*
Faculty of Civil Engineering
Delft University of Technology
P. O. Box 5048825
2600 GA Delft NETHERLANDS

Roger Lovegrove
Building Research Establishment
Garston Watford WD2 7JR ENGLAND

M

Henrik O. Madsen *(Author and Lecturer)*
Det Norske Veritas, Danmark
Nyhav 16
1051 Copenhagen K DENMARK

Catherine Marx
Southern Forest Products Association
P. O. Box 52468
New Orleans, LA 70152 USA

Thomas E. McLain
Virginia Polytechnic Inst. and State Univ.
Brooks Forest Products Center
Blacksburg, VA 24061-0503 USA

John Mentis *(Co-sponsor)*
National Forest Products Association
1250 Connecticut Ave., NW
Washington, D.C. 20036 USA

Joseph Murphy
Structural Reliability Consultants, Inc.
P. O. Box 56164
Madison, WI 53705-9464 USA

Raphael N. Mutuku
Civil Engineering Department
Univeristy of Nairobi
P. O. Box 30197
Nairobi KENYA

N

Takashi Nakai
Forestry and Forest Products Research Institute
Ministry of Agriculture, Forestry and Fisheries
P. O. Box 2
Ushiku, Ibaraki
300-12 JAPAN

Lauge Fuglsang Nielsen *(Author and Lecturer)*
Building Materials Laboratory
Technical University of Denmark
Building 118
DK-2800 Lyngby DENMARK

O

Lars Ostlund
Dept. of Structural Engineering
Lund Institute of Technology
P. O. Box 118
S-22100 Lund SWEDEN

R

Rudiger Rackwitz *(Author and Lecturer)*
Technical University of Munchen
Arcisstrasse 21
D-8000 Munchen 2 GERMANY

David Rosowsky
School of Civil Engineering
Purdue University
West Lafayette, IN 47907 USA

LIST OF ATTENDEES (continued)

R (continued)

Frederic Rouger
Assistant Professor
Universite de Technologie de Compiegne
Departement de Genie Mecanique
Division Modeles Numeriques en Mecanique
BP 649
60206 Compiegne Cedex FRANCE

S

John Scalzi *(Co-sponsor)*
National Science Foundation
1800 G Street, NW
Washington, D.C. 20550 USA

T

Juhani Taipale
Laboratory of Structural Engineering
Technical Research Centre of Finland
Kemistintie 3 02150 Espoo FINLAND

Sven Thelandersson
Department of Structural Engineering
Lund University
Box 118
S-221 00 Lund SWEDEN

Robert Tichy
Robert J. Tichy and Associates
2012 S. 314th, Suite 233
Federal Way, WA 98003 USA

U

Luca Uziella
Professor of Wood Technology
Universita degli Studi de Firenze
via S. Bonquentura, 13-I 50145
Firenze ITALY

V

Erol Varoglu
Forintek Canada Corp.
2665 East Mall
Vancouver, B.C. V6T 1X5 CANADA

W

Bryan Walford
Forest Research Institute
Private Bag 3020
Rotorua NEW ZEALAND

Thomas G. Willamson
American Institute of Timber Construction
11818 SE Mill Plain Blvd., Suite 415
Vancouver, WA 98684-5092 USA

Z

John Zahn
Forest Products Laboratory
One Gifford Pinchot Drive
Madison, WI 53705 USA

NATO ARW "Reliability-Based Design of Engineered Wood Structures"

Group Assignments

Director	:	Jozsef Bodig	U.S.A.
Keynote Speaker	:	Bruce Ellingwood	U.S.A.

GROUP A: Fundamentals of Reliability Assessment

Organizing Committee Member	:		
Speaker	:	Henrik Madsen	Denmark
Group Leader	:	Sven Thelandersson	Sweden
Rapporteur	:	Frederic Rouger	France
Participants	:	Bruce Ellingwood	U.S.A.
		Zsolt Kovacs	Hungary
		John Zahn	U.S.A.

GROUP B: Reliability Assessment of Multi-Member Structures

Organizing Committee Member	:		
Speaker	:	Rudiger Rackwitz	Germany
Group Leader	:	Marita Kersken-Bradley	Germany
Rapporteur	:	Joe Murphy	U.S.A.
Participants	:	Marvin Criswell	U.S.A.
		Lars Ostlund	Sweden
		David Rosowsky	U.S.A.
		Erol Varoglu	Canada

GROUP C: Material Resistance Considerations

Organizing Committee Member	:	Dave Barrett	Canada
Speaker	:	Ricardo Foschi	Canada
Group Leader	:	Bob Tichy	U.S.A.
Rapporteur	:	Roger Lovegrove	U.K.
Participants	:	Jean-Pierre Biger	France
		Ario Ceccotti	Italy
		Vijaya Gopu	U.S.A.
		Cathy Marx	U.S.A.
		Takashi Nakai	Japan
		Lauge Nielson	Denmark
		Juhani Taipale	Finland

GROUP D: Formulation of International Harmonized Codes

Organizing Committee Member	:	Bob Leicester	Australia
Speaker	:	Bob Leicester	Australia
Group Leader	:	Geoff Boughton	Australia
Rapporteur	:	Bryan Walford	New Zealand
Participants	:	Erik Aasheim	Norway
		Ron Anthony	U.S.A.
		Ad J. M. Leijten	Netherlands
		Tom McLain	U.S.A.
		Raphael Mutuku	Kenya
		Tom Williamson	U.S.A.

INDEX

E

I

G

I

L

M

N

O

P

Q

R

T

U

MIX
Papier aus verantwortungsvollen Quellen
Paper from responsible sources
FSC® C105338

If you have any concerns about our products,
you can contact us on
ProductSafety@springernature.com

In case Publisher is established outside the EU,
the EU authorized representative is:
Springer Nature Customer Service Center GmbH
Europaplatz 3, 69115 Heidelberg, Germany

Printed by Libri Plureos GmbH
in Hamburg, Germany